National Water Security

Mustapha Besbes · Jamel Chahed
Abdelkader Hamdane

National Water Security

Case Study of an Arid Country: Tunisia

Springer

Mustapha Besbes
University of Tunis El Manar, National Engineering School of Tunis
Tunis
Tunisia

Abdelkader Hamdane
University of Carthage, National Agronomic Institute of Tunis
Tunis
Tunisia

Jamel Chahed
University of Tunis El Manar, National Engineering School of Tunis
Tunis
Tunisia

ISBN 978-3-030-09247-4   ISBN 978-3-319-75499-4 (eBook)
https://doi.org/10.1007/978-3-319-75499-4

Printed on acid-free paper

This Springer imprint is published by the registered company Springer International Publishing AG part of Springer Nature
The registered company address is: Gewerbestrasse 11, 6330 Cham, Switzerland

Foreword

Little by little, a general awareness is spreading: the "water problem" is becoming one of the questions of increasing urgency that must be answered within the coming years by the more than seven billion inhabitants living on our planet today or by the nine and a half billion who will most likely be living here in 2050, or more than eleven billion in 2100. There are three main reasons for this: the seemingly unstoppable demographic growth, although the rhythm of this growth appears to be slowing down in some countries, but shows no signs of doing so in others, particularly in sub-Saharan Africa; the rising demand for water due to economic development and increasing food consumption, as the major proportion of the water consumption is linked to food production. The changes in diets and food habits promise an enormous increase in the demand for agricultural water which represents, by far, the largest portion of our water needs, and is today on average around 800 m^3/y per inhabitant, in the poorest countries, but over 2000 m^3/y in the richest, or most wasteful, ones who unfortunately set a bad example for the rest of the world; the third reason is the predicted climate change which is likely to cause a highly altered rainfall distribution, in particular a probable further desertification in the current arid zones, increased rainfall in the northern latitudes, and a greater intensity of extreme events. To these three threats to the water quantity, one must add the continuous degradation of the water quality due to a general spread of pollutants by humans throughout the environment, as well as to contamination by salt and other natural elements, such as arsenic and fluoride, indirectly set in motion by human activity. All the countries in the World, whether developed or developing, are faced with these problems.

The present situation is not sustainable. We must change our practices. But how do we go about it? This book offers a wealth of information and thoughtful discussions of these questions, starting at the world scale and then concentrating on the case of Tunisia, which is probably the country in the arid zone that knows the most about its water resources and needs, after 50 years of practicing a determined policy of development and resource management based on an increasingly precise understanding of the water cycle and of the available resources. The arid zone is characterized by a scarcity of resources, high inter-annual variability, very often strong salinity, and general overexploitation of the groundwater resources. Tunisia is a textbook case that may serve as a model for many countries in the World, for its detailed water resources evaluation and monitoring, and for its water management techniques and its water governance, because of the successes as well as the failures it has encountered.

Before treating the water problem, one must first define its scope. This book starts very appropriately by a general presentation of water problems on the global scale and the most recent thinking in water resources and consumption, distinguishing between blue water, which flows in rivers and aquifers and can be withdrawn, green water which is stored in the upper layers of the soil and can only be used by the vegetation through root uptake. The green water, the one used in rainfed agriculture, is usually excluded from the water balance and from the water scarcity question, although it represents an essential part of the resources that we use. The authors explain the concept of "water footprint", which quantifies the amounts of blue or green water needed to produce what we consume, agricultural and industrial products as well as domestic water. In the light of this investigation, an exhaustive examination is undertaken of the demand–resource balance in Tunisia beginning with a backward perspective of nearly 100 years, a history of development and regulation of the water supply, followed by a discussion with a forward perspective, until 2050. Here, the authors propose scenarios for the offer as well as for the demand in order to arrive at "water security" for Tunisia. At present, around 39% of the water consumed in the country is imported, what is called "virtual water" and represents the water contained in products bought abroad, mainly cereals, for which the water needed to grow them has been estimated. "Hard water security" is then understood as the minimum local food production that must be guaranteed for the survival of the population in a time of crisis when food importation is impossible. Water security is thus strongly linked to food security. Is it reasonable to exceed 39% of virtual water import in the national water balance? What to do when the forecasted water balance in 2050 may reach 49% of virtual water import, or even 64% during dry years? This question leads to the definition of the important notion of *Water Dependency Index*, which is the ratio of the imported virtual water minus the exported one, divided by the water equivalent of the food consumption. Following the evolution of this index through time is very instructive.

Water security also includes protection of the resources and their infrastructure from pollution, accidents, malicious acts, of whatever origin, and anticipation of extreme hydrologic events. All these subjects are treated with clarity and competence.

An important, and very original, part of the book is dedicated to a detailed scrutiny of the quantity of water consumed by Tunisian agriculture for food production by rainfed and irrigated crops, respectively. These, almost never estimated, figures lead to a surprising conclusion: the major part of Tunisian food production is provided by rainfed agriculture; irrigated crops are far behind although they have a greater added value and can be exported, e.g., citrus fruit. But olive oil, produced by rainfed agriculture, is also a very important part of food exports. Tunisia's food exports financially offset its virtual water import, which is mostly cereals. At a time when the resources available for irrigation are reaching their limit, the authors show that investments to improve the efficiency of the hydraulic systems must continue but that the main efforts must now be concentrated on optimizing the yield of rainfed agriculture, with better agronomic practices and soil water conservation. This a priori surprising conclusion is very carefully reasoned and discussed.

A very interesting concept is that of food storage versus water storage. In a country subject to strong inter-annual rainfall variability, there is a need to balance years with rainfall deficits with years of above-average rainfall. But storing water in a dam is very expensive, subject to evaporation losses and environmentally questionable. A simple alternative is to store food, which in fact serves the same purpose, when water is used to produce this food. Tunisia, for instance, has grain stores of over half a million metric tons and aims to store, in the near future, more than 6 months' worth of grain consumption, which is equivalent to storing 2.4 km^3 of water. The cost is low, \$42 per metric ton in Tunisia, and even \$15 in South Korea. Russia has revealed earlier this year its plans to build very large food stores valued at billions of dollars. Similarly, Tunisia stores around 365,000 metric tons of olive oil, i.e., twice the annual production and equivalent to 3 km^3 of water.

The book deals with other essential questions, such as the importance that must be given to desalination of brackish water or seawater, the use of treated wastewater, saving water in all types of activities, what share of available water should be attributed to natural ecosystems such as wet zones, water supply in urban and rural environments, wastewater disposal, etc. Furthermore, major attention is given to governance, including the necessary participation of the public in decision-making and local administration of the resources. These questions are fundamental to good water administration. Here, Tunisia is again a textbook example: it has a strong central power and a very active, competent administration who has nevertheless attempted to delegate management to Associations of Water Users of which some have worked very well, others less so. The revolution in 2011 left its mark due to the weakening of the authority of the State, and excesses were committed which increased the vulnerability of an already fragile resource, for example, illegal connections to the main water pipes or undeclared drillings in zones protected because of excessive withdrawals. The difficult balance between centralized authority, delegation of power, management shared with the users, and conflict solving is discussed exhaustively. However, a prerequisite to shared management is, according to the authors, the education and information of the public about the hydrological realities and water management. The place given to research in helping to shed light on policy choices is also discussed.

This work, written in a very pedagogic and extremely clear language, is teeming with data and information that are thoughtful, scientific, and pertinent and shed new light on the thorny issue of water. Everybody who is interested in water will find it to be a source of original ideas that can be applied to the choices that must be made in the future, in Tunisia, and elsewhere because the example of Tunisia is used only as a basis to illustrate a universally prevalent situation. I congratulate the authors on their remarkable achievement and warmly recommend this book to everybody interested in water, and particularly to students learning about water resources and management.

Paris, France
November 20th, 2017

Ghislain de Marsily
Emeritus professor
University Pierre et Marie Curie, Paris
Member of the French Academy of Sciences, Foreign
Member of the US National Academy of Engineering

Preface

This book is the direct result of a decades-long reflection on the management of water scarcity, and on the need, in Tunisia and thereafter in some arid countries, to apprehend water resources as a whole; the original and essential resource being that of precipitation. With the contribution of the authors, the initial product was the publication in 2002 by the Tunisian Institute of Strategic Studies (ITES) of the report on "The future of water, a new challenge for Tunisia". This report developed more thoroughly the holistic view of water management, across the integration of blue water, green water, and virtual water components. In 2003, the ITES associated the authors to the prospective "Tunisia 2030", establishing that the scenario "Integral Water Vision", could open up prospects for controlling long-term water dependency of the country, in total disruption with trend scenarios based only on blue water.

Reported in publications and presentations, the results and methodologies developed have made it possible to disseminate the Tunisian experience among the scientific community. The findings for the Tunisian case have met with the support of the most active researchers on the subject in the world. In this regard, we would like to pay tribute to the experts and eminent scientists for their encouragement and their interest in our work: John Anthony Allan, Akissa Bahri, Ghislain de Marsily, Arjen Hoekstra, Ramon Llamas and Daniel Zimmer.

Meanwhile, the holistic water vision and the perspectives it may open in arid context have made their way in Tunisia, both in scientific circles and among administrative bodies with obvious impacts on the water resource and on the legislative framework for its management. In 2010, the ITES commissioned the authors a new study on the "National Water Balance", which was conducted using the same principles, those of the holistic vision of water resources. In the same year, the Ministry of Agriculture asked the authors to contribute to the preparation of the new "Water Code", the new version of which introduces agricultural soil water as an integral component of water resources, and reaffirms the principle of optimum water efficiency by extending it to all forms of resources, including green water.

The authors have gathered all their respective experiences of water problems in Tunisia in a French book, published in 2014 by the Harmattan Editions in Paris under the title "Water Security of Tunisia, managing water under scarcity." [« Sécurité Hydrique de la Tunisie : Gérer l'eau en condition de pénurie »]. The almost total mobilization of blue water resources marked a turning point for Tunisia: this book shows how, therefore, change in the water paradigm had become a necessity, and provides elements for new policies adapted to real water potentialities.

The opportunity that was graciously offered by Gunilla de Marsily to translate into English a condensed version presented at an English-speaking event encouraged us to write this entirely English version in a broader perspective. Denis Pryen, the Director of Editions "L'Harmattan", has authorized us to translate the French version; many thanks to him. Last, we are grateful to Nabil Khelifi, our Springer Senior Editor, for agreeing to review, evaluate, and publish our English manuscript. The observations made by the two anonymous reviewers, responsible for evaluating the project, allowed us to very significantly improve the content of the manuscript. We thank them all.

This book covers about two-thirds of the French edition content, which has been revised and updated, particularly with regard to new data. While maintaining the original thematic contents, the English version develops, makes explicit, and applies to Tunisia recent concepts, formalisms, and tools, particularly the emerging water–agriculture–food–environment nexus whose relevance becomes even more evident with the widespread application of the water footprint and virtual water concepts. Building on the momentum generated by the debate on water and food security, the book develops a conceptual model for National Water Security for an arid country "Tunisia" with the potential objective of transposing, extending, or adapting it to other water-scarce countries and regions whenever possible.

Tunis, Tunisia
November 24th, 2017

Mustapha Besbes
Jamel Chahed
Abdelkader Hamdane

About the Book

This book shows how the change of water paradigm has become urgent, and provides evidence for the need of new policies that expand water balance to green and virtual water. The issue of National Water Security concerns drinking water supply but also food safety, linked to agricultural policy. Both rainfed and irrigated agriculture play complementary roles in food security, and the water issue implies a holistic view of water resources. This view constitutes the book's backstory. The reader will find original ideas that can be applied everywhere because the example of Tunisia is typically a basis to illustrate a universally prevalent situation. The book deals with other important issues: desalination, wastewater recycling, water quality, groundwater overdraft, water savings, governance, knowledge valuing, education, information: upgrading the whole water systems for the future implies emancipation of the whole society.

Contents

About the Authors

Mustapha Besbes is hydrogeologist, emeritus Professor at the National Engineering School of Tunis, University of Tunis El Manar. He received the PhD in Hydrology (1967) and the Doctorat es sciences from the University Pierre & Marie Curie in Paris (1978). Pr Besbes was elected Foreign Member of the French Academy of sciences in 2009, and member of the Tunisian Academy of Sciences, Letters and Arts in 2013. World leader in arid zone hydrology, he coordinated the NWSAS (North Western Sahara Aquifer System) modeling process in the 2000s. He has published books and articles in hydrogeology and water management. He is chairman, reviewer or associate editor of journals in hydrological sciences. Mustapha Besbes is member or head of numerous scientific commissions, councils, boards and task forces.

Jamel Chahed is a hydraulics engineer and Doctor es sciences, Professor at the National Engineering School of Tunis, University of Tunis El Manar. His work focuses on transfers in industrial and environmental fluid systems, atmospheric dispersion, water governance and water management. His research has appeared as articles, books, book chapters, conference acts in specialized literature such as hydrology, water resources management, hydraulics, turbulence and transfers, environmental engineering. He actively contributed to the development of long-term vision for water security with a thematic focus on holistic water balance assessment. He has been involved in number of national and international academic research and technical programs. As visiting professor, associate researcher or guest speaker he joined number of international research teams at academic institutes, research laboratories and R&D centres.

Abdelkader Hamdane is graduated from Institut National Agronomique de Paris (1971), and Ecole Nationale du Génie Rural, des Eaux et Forêts de Paris (1973). He is general engineer, formerly General Director of the Medjerda Valley Development Company, General Director of Rural Engineering and Water Management Department of the Ministry of Agriculture and Water Resources of Tunisia. He has been member of several public companies boards operating in the field of water, and led several assessment missions in scientific and technical research institutions. He was a short-term professor of irrigation and drainage at the National Agronomic Institute of Tunis and the National Engineering School of Tunis. He is currently a Scientific Advisor at the National Agronomic Institute of Tunis, University of Carthage, involved in many national and international research programs, and consultant on water management in international organizations.

Acronyms and Abbreviations

ANPE	Agence Nationale de Protection de l'Environnement (Environment Protection National Agency)
BIRH	Bureau d'Inventaire des Ressources Hydrauliques (Office for Water Ressouces Inventory)
CRDA	Commissariat Régional au Développement Agricole (Regional Commission for Agricultural Development)
DGACTA	Direction Générale de l'Aménagement et de la Conservation des Terres Agricoles (General Directorate of Agricultural Land Development and Conservation)
DGBGTH	Direction Générale des Barrages et des Grands Travaux Hydrauliques (General Directorate of Dams and Large-scale Hydraulic Works)
DGGREE	Direction Générale du Génie Rural et de l'Exploitation des Eaux (General Directorate of Rural Engineering and Water Exploitation)
DGRE	Direction Générale des Ressources en Eau (General Directorate for Water Resources)
DHMPE	Direction de l'Hygiène du Milieu et de la Protection de l'Environnement (Directorate of Environmental Hygiene and Environment Protection)
DPH	Domaine Public Hydraulique (Public Water Domain)
DTP	Direction des Travaux Publics (Directorate for Public Works)
ET	Evapotranspiration
GDA	Groupement de Développement Agricole (Agricultural Development Association)
GWP	Global Water Partnership
HTNPP	Human appropriation of Terrestrial Net Primary Production
INM	Institut National de la Météorologie (Meteorological National Institute)

INNORPI	Institut National de la Normalisation et de la Propriété Industrielle (National Institute of Standardization and Industrial Property)
INRGREF	Institut National de la Recherche en Génie Rural, des Eaux et Forêts (National Research Institute for Rural Engineering, Water and Forestry)
INS	Institut National des Statistiques (National Institute of Statistics)
IPCC	Intergovernmental Panel on Climate Change
IWMI	International Water Management Institute
IWRM	Integrated Water Resources Management
MGD	Millennium Development Goals
NIDIS	National Integrated Drought Information System
NOAA	National Oceanic and Atmospheric Administration
NWIS	National Water Information System
NWSAS	North Western Sahara Aquifer System; (SASS: Système Aquifère du Sahara Septentrional)
OECD	Organisation for Economic Co-operation and Development
ONAS	Office National de l'Assainissement (National Sanitation Office)
ONTH	Office National du Thermalisme (National Thermalism Office)
ORSEC	Organisation des Secours (Organization of the Help)
PAUE	Périmètre d'Aménagement et d'Utilisation des Eaux (DWUP: Development and Water Use Perimeter)
PDEC	Plan d'utilisation des Eaux du Centre (Water Use Plan for the Center)
PDEN	Plan d'utilisation des Eaux du Nord (Water Use Plan for the North)
PDES	Plan d'utilisation des Eaux du Sud (Water Use Plan for the South)
PET	Potential Evapotranspiration
PISEAU	Projet d'Investissement dans le Secteur de l'Eau (Water Sector Investment Project)
PNEE	Programme National d'Economie d'Eau (or NWSP: National Water Saving Program)
RW	Recycled Water
SECADENORD	Société d'Exploitation du Canal et des Adductions des Eaux du Nord (Northern Water Canal and Adductions Exploitation Company)
SINEAU	Système National d'Information sur l'Eau (National Water Information System)
SONEDE	Société Nationale d'Exploitation et de Distribution des Eaux (National Water Exploitation and Distribution Company)
SSO	Sahara and Sahel Observatory; (OSS: Observatoire du Sahara et du Sahel)

STEG	Société Tunisienne de l'Electricité et du Gaz (Tunisian Company of Electricity and Gas)
TNPP	Terrestrial Net Primary Production
TWR	Treated Wastewater Reuse
WDI	Water Dependency Index
WF	Water Footprint
WHO	World Health Organization
WPI	Water Poverty Index
WWC	World Water Council

List of Figures

List of Tables

General Introduction

At the end of five decades of intense mobilization of its material and intellectual capacities, Tunisia has managed to identify and exploit, for its development, the greater part of its natural water resources: it has thus endowed itself with a water infrastructure that allows to provide safe water for almost all urban and rural populations and develop the irrigation sector, which contributes to more than a third of agricultural production. These water achievements also contribute to greatly mitigating climatic hazards as well as flood damage. The implementation of the national water system has also contributed to the emergence of large-scale capacities in the scientific and technical fields related to water, and the need for the establishment of economic systems and original social organizations to try to control at best the scarcity of the resource.

Despite all these advances, but also because of a strong growth demand, signs of reaching water resources limits have been clearly identified since the beginning of the 1990s, and the water sector experiences profound transformations, marked by a very high level of water resources exploitation: use of the best sites for large dams, groundwater overexploitation, increased use of unconventional water resources, salinization of soils in irrigated areas, development of water drinking and irrigation at the expense of groundwater recharge, reduced capacity of dams, etc. Simultaneously, the socioeconomic conditions have changed with the higher living standards of the Tunisian population and the attendant greater demand for water quality and supply security, stronger and effective users' participation in local water management, increased needs for information and training, and more and more citizen involvement in water policy.

Potentially, water may become a limiting factor in socioeconomic development of the country and this fact ought to provoke a thorough review of its management principles: how to move from supply management to demand management? How, in a context of increasing competition for water, can we reconcile different uses, anticipate conflicts, ensure resource conservation, and control risks? How to combine the need to respect the unique nature of the hydrological resource with a centralized organization of water administration and management? What governance model should we adopt? What role should we assign to the private sector and to community management? How far should we go in the use of economic tools?

The time is past when we could easily meet additional demand by putting new supplies into production. With the water demand continual increases prospects, unsustainable tensions should be exerted on a finite resource, already well underway, in quantity and quality. This issue, which concerns all countries on the southern and eastern Mediterranean, has led a number of experts to question the future of the world various regions in terms of allocation of water resources and on geostrategic implications of possible serious shortages. It happens that in this unenviable competition, Tunisia is ranked among the last of the planet, a situation that is emphasized by the relevant international water actors [F.A.O, World Bank, Blue Plan, United Nations]. However, the Tunisian today has a level of "Water Comfort", or mobilized conventional freshwater available per capita, which has never been equaled in the country's history. And if it happens to farmers to undergo irrigation water ration in drought years, they compensate most often by overexploiting their free access groundwater. As for urban population (68%), it does not seem to be prepared for a situation of permanent water stress or chronic water shortage. As a result, managing the resource from now on will also be more about driving new approaches to supply and demand, and new users behaviors.

The analysis of these water fundamental issues tries to find ways of thinking about essential principles, concepts, and instruments on which water policies should be based in order to make the best use of limited resources. It is a question of defining the conditions of a long-term water security adapted to arid countries which guarantee a water and soil resources sustainable development while preserving the wealth and the well-being that they have so far accompanied. In particular, the question asks about the socioeconomic support needed to develop new policies that can prepare, for the future, the most important measures to be taken, on the various technical, economic, cognitive, legislative, and institutional levels. Ultimately, this approach should lead to the definition of a conceptual model of National Water Security which, through the Tunisian example, develops a vision that can be transposed to other arid countries.

The issue of water security obviously concerns the security of the drinking water supply but also and particularly, food safety, inextricably linked to agricultural policy and food balance. In the world, over 70% of exploited water resources, in Tunisia almost 80%, are allocated to irrigation. Both rainfed and irrigated agriculture play essential and complementary roles in ensuring food security. Irrigated agriculture ensures the increase and stabilization of local agricultural production and plays a key role in promoting rural development. As for rainfed agriculture, it contributes significantly to food security and plays a crucial role in the agricultural trade balance. Given this relationship between water security and food security, the approach to the water issue passes necessarily by a holistic view that considers the multiple stress factors likely to affect the water supply.

Water security can be defined simply as sustainable access to water in sufficient quantity and of acceptable quality, but there is an urgent need for universal tools to characterize and measure the level of water security. Many indicators are used for this purpose, most of which try to evaluate the anthropogenic pressure on the resource, but do not refer to water governance, including the existence of infrastructures that provide effective access to water. In this respect, knowledge and information on the resources will be key elements for assessing the state of hydrological environments and pressures, limiting and preventing the consequences of threats and risks, and involving institutions and the public to ensure responsible participation.

Other dimensions combine to frame the water security concept. They concern "hard security" and involve both internal and external security components. Internally, some countries place special importance on the protection of water resources and installations against possible terrorist acts, while others are introducing appropriate equipment to prevent emergencies in the drinking water supply of large metropolises in the event of a serious crisis. Externally, water security includes the cross-border dimension: one hundred and fifty countries around the world share a portion of their water resources with one or more neighboring states.

In Tunisia, the strengthening of long-term water security will depend on the country's ability to find adequate and appropriate solutions to satisfy ever increasing water quantity needs and quality requirements. The importance allotted to the various solutions and the space they will be allowed in strategies and water resource development programs must be assessed in the light of resource availability, quantitative and qualitative resource sustainability, optimization of different uses of various types of water resources, and the ability to adapt and implement institutional reforms to accompany the new water policies, including the capacity to make all water stakeholders join in the reforms.

The present book is organized around four themes at the heart of water issues: (i) First, the knowledge and control of the national water balance the framework of the debate on National Water Security, developed through a comprehensive review of the evolution of ideas and concepts that, in less than 30 years, have gone from a spatially delimited national vision of water stress to an open vision, which provides the globalization of water resources with the generalization of virtual water and water footprint paradigms; (ii) then the question of the relationship between water security and food security, which is based on a holistic view; (iii) emphasize the value of water through demand management, by water resources optimal use and control of the hydraulic systems efficiency; and (iv) improve the security of water supply and achieve better water governance, focusing on the major issues that are currently debated by experts but also by the public and which will determine some of the future challenges. Each of these questions is the subject of a chapter.

It seemed, however, necessary to treat these fundamental themes from four perspectives: (i) the framework of the debate on National Water Security, developed through a comprehensive review of the evolution of ideas and concepts that, in less than 30 years, have gone from a spatially limited national vision of water stress to an open vision, which provides the globalization of water resources with the generalization of virtual water and water footprint paradigms; (ii) a geographic and international perspective, as an overview of the major water issues in the World, especially those related to arid and water scarce countries; (iii) a historical and analytical perspective of water policies in Tunisia; and (iv) the future perspective, with an emphasis on the concept of national water dependency, the outlook for blue and green water resources, a national green water resource estimation, and perspectives on long-term holistic water balances.

Chapter 1
The World Water Issues

Contents

1.1 Introduction: Could Humanity Undergo Water Shortages?

Continental precipitation forms the common freshwater resource[1]. This capital, estimated at 110,000 billion m^3 per year (km^3/year), is unevenly distributed: arid regions that receive little, suffer permanent water stress; it is called physical shortage. But lack of water also has an economic origin. This is the case in many well-watered areas where insufficient infrastructure produces water shortages. The result is that in 2017, one in ten people in the world has no access to safe drinking water.

At the global level, 64% of the rainfall is taken up by Evapotranspiration: 57% by forests, grasslands, wetlands and only 7% by cultivated land. The remaining 36% generate flows: in rivers and groundwater where irrigation, practiced on 300 million hectares (Mha) represents the major withdrawals (70% or 2800 km^3/year). Cities, industries and hydropower use the rest (1200 km^3). It should be mentioned, however,

[1]A french version of this text has been published by the daily newspaper Le Figaro, Paris, on January 19, 2012 in the section dedicated to the French Academy of Sciences.

M. Besbes et al., *National Water Security*,
https://doi.org/10.1007/978-3-319-75499-4_1

that quantities of water consumed by rainfed and irrigated agriculture to meet the human food requirements represent 95% of our total water demand; the rest is used for drinking, domestic water and industries (Besbes 2012).

Many countries do not produce all their food but import part-sometimes quite large-of it as "virtual water" (water embedded in products bought abroad). The total flow of this virtual water (1600 km^3/year) reflects the "globalization" of water resources. Countries that are highly dependent on virtual water do not experience serious stress if they have sufficient economic power. In contrast, obliged to reduce their food bills, the less wealthy must necessarily optimize these flows by developing local capacity for production and storage of strategic food products.

Because of population growth and improved standard of living, global food demand could double by 2050. However, water and soil resources are already severely depleted by irrigation with serious environmental impacts: groundwater overexploitation (22 million wells in India), soil salinization (20 million hectares affected in the world), river artificialization, shrinking wetlands and water-quality degradation. To make matters worse, global warming is expected to exacerbate the situation. How, under these conditions, do we continue to meet the growing needs for irrigation that produces 40% of the world's food on only 20% of its arable lands? Water withdrawals and irrigated areas are therefore expected to grow strongly, especially in countries that already suffer from water stress. Without major changes, this will be difficult. Radical reforms will have to be implemented: more efficient irrigation, drought-tolerant plants, fair pricing, and support for farmers. The aim is to produce more with less water, while preserving the ecosystems.

Rainfed agriculture covers 1300 million ha, globally 80% of cultivated areas (90–95% in the Maghreb and sub-Saharan Africa), and produces 60% of the food on the planet. In some parts of Europe and North America, it is possible to develop high efficient rainfed agriculture. But in many arid regions, agricultural yields undergo dramatic changes due to weather conditions. In these situations, water scarce countries should strengthen the capacity of traditional drought adaptation: terracing, anti erosive facilities, spate irrigation and groundwater recharge, drought-resistant plant varieties.

Industry is the activity where water demand is set to increase in the developing areas of the world as a result of off shoring and equipment of emerging countries. To meet these new requirements, alternative resources will have to be developed. In 2010, recycling concerned 7 km^3/year or 4% of the collected and treated urban wastewater but it still offers an interesting perspective for agriculture and industry; desalination of brackish water and seawater produced 8 km^3/year which represent 0.2% of the freshwater consumed worldwide. As production costs are becoming increasingly competitive, desalination could double in ten years, even if its energy cost and environmental impact (waste brines) are still high.

Another crucial issue: urban drinking-water supply has always been a top priority. The world was 70% rural in 1950, but it will be 70% urban by 2050. The urban population explosion looming in developing countries raises fears of new shortages: in these countries, cities will have more than 5 billion inhabitants in 2050. There is therefore a need to continue to go further for better quality drinking water and make greater use of desalination, even if this increases the water bill for communities which sometimes lack the necessary means. But other solutions exist that allow the urban water cycle to operate in a coordinated and efficient manner, (Bahri 2012). In any event, the need for international cooperation is essential, without it, the generous resolution 64/292 of the United Nations in July 2010 recognizing the right to water for all will be ignored.

This first chapter, which attempts to familiarize the reader with some concepts and debates that dominate the water issue at the global scale, includes four parts:

- The first part (Sect. 1.2) develops the world water balance. It outlines the contributions made to the development of understanding the renewable freshwater resources on Earth, and the global water balance, as well as freshwater withdrawals, impact of human actions on the hydrological cycle. The concept of water footprint applied at global scale makes it possible to specify the respective places occupied by the blue water, green water, and virtual water concepts in the world budget.
- The second part (Sect. 1.3) presents a focus on water balance of arid countries. While at the global scale, total flow on the continents represents 36% of precipitation and evapotranspiration 64%, in arid countries, the total flow is limited to 10% and evapotranspiration reaches 90% of rainwater resources. Therefore, while they occupy 18% of the continents, arid countries produce only 1.3% of the internal renewable water resources of the world. This number figure should be compared with the population of arid countries, which represents 8% of the world. Another consequence: arid countries withdraw 86% of their renewable water resources, which raises the question of the sustainability of such a pace of withdrawal.
- The third part (Sect. 1.4) deals with the impacts of the anthropogenic activities on water resources. These are particularly acute in most semi-arid and arid countries where strategies of massive resource mobilization and transfers are faced with problems related to water scarcity and competing demands for water. The perspective of climate change due to global warming may exacerbate an already critical situation in some countries experiencing water stress: increased mobilization of water resources, water quality deterioration, groundwater overexploitation, impacts of irrigation on salinization of soils and aquifers, impacts of dam reservoirs on groundwater recharge.
- The fourth (Sect. 1.5) part presents the main principles, concepts and instruments of water policies. Among the basic principles on which water policies in the world are based, should be mentioned, the user pays and polluter pays principles, the precautionary principle and the principle of subsidiarity. As regards the instruments of water policies, and among the most important, the different management approaches are presented: Integrated Water Resources Management (IWRM), water demand management, watershed-based water management, participatory management. Mention is also made of aspects relating to the economic dimension of water (value and pricing) as well as water data and water information systems.

1.2 The World Water Balance

1.2.1 The Water Cycle and Global Water Budget

Fresh water is an integral part of the terrestrial water circulatory system constituted by three stages: evaporation, condensation, precipitation, and called the "water cycle". The water cycle shows the existing flows between major water reservoirs of our planet: oceans, glaciers, atmosphere, lakes and marshes, rivers, soils, aquifers. These flows are powered by solar energy, causing water evaporation and the exchange cycle.

Under solar radiation, ocean water [97% of the world water reserves] and inland water bodies evaporate. The evaporated water is pure and free of salts. The estimates of renewable freshwater resources fed by rainfall on the continents vary between 108,000 and 119,000 km^3/year (Table 1.1), on a total area of 149 million km^2, an equivalent of a rain water depth between 725 and 800 mm/year.

Rainfall on the continents produces the various elements of the continental hydrologic cycle. A certain amount of water is temporarily stored on the land surface into lakes and glaciers: the superficial stock. Another part of rainfall flows over the ground; drained into watercourses, it forms the runoff. When the land properties (porosity, permeability) are favorable, water penetrates into the ground by infiltration. Under the ground surface, the pores of the formations contain both air and water. This is the unsaturated zone or vadose zone which still contain significant amounts of water, abstracted by the vegetation root system and which feed plant evapotranspiration. These quantities are soil moisture, or soil water reserve. The excess soil moisture is conveyed downward by gravity to the groundwater reserves. These flow underground to, in the end, reach the surface drainage network, the system of surface water as springs, or leak into the sea.

The Russian School of Hydrology, including researchers from the St. Petersburg Hydrology State Institute (Lvovic and Shiklomanov) gave the first attempts on world water balance assessment. This vocation is due to the size of the former Soviet Union territories and the early use of hydrological mapping processes, ancestor of the Geographic Information Systems. Through these processes and numerous rainfall observations and runoff data, the Russian hydrologists were able to offer the first models quantifying the terms of the world water balance. Table 1.1 shows the various successive estimates of flows circulating in the hydrosphere, from which one can observe the following: (i) The knowledge gained over the past 50 years have not upset the orders of magnitude of the global water balance; (ii) Fresh water resources of the world (total precipitation) are of the order of 550,000 km^3/year, 20% of which represent the continental rainfall resources.

Table 1.2 presents estimates of the water stock contained in the various reservoirs of the hydrosphere. It may be noted that out of a total of 1380 million km^3, almost 97% are in the form of salt water in the oceans, seas, salt lakes and salt groundwater. All the fresh water reserves of the world represent 3% of the volume of water on earth, and among these reserves, 69.5% are retained as ice. The fresh groundwater accounts for 30%, lakes 0.01%, soil water reserve 0.001% as well as water in the

Table 1.1 The successive world water balance estimates

Components	L'vovic (1968)		Korzoun Sokolov (1978)	Shiklomanov (1993, 1998)		Chahine (1992)	Harvey (2000)	de Marsily (2009)	
Units	1000 km^3/year	mm/year	1000 km^3/year	1000 km^3/year	mm/year	1000 km^3/year	1000 km^3/year	1000 km^3/year	%
Rainfall over oceans	411.6	1143	458	458	1272	398	458		
Continental water input	36.4	101	47	44.8	124	37	47		
Evaporation in oceans	448	1244	505	502	1397	435	505		
Rainfall over continents	108	728	119	119	799	108	119	113	100
Total flow	36.4	244	47	44.8	301	37		39.6	35
Of which glaciers								3.4	3
Of which total surface flow				42.7	287		45.3	34	30
Of which groundwater undrained by river				2.1	14		1.2	2.3	2
Evaporation-Evapotranspiration over continents	72	483	72	74.2	498	71	72	73.5	65
Rainfall over the globe	520	1020	577	577	1134	506	577		
Evaporation over the globe	520	1020	577	577	1134	506	577		

Table 1.2 Distribution of water on Earth

Medium	L'vovic (1968)	Shiklomanov Sokolov (1983)	Gleick (1996)	Harvey (2000)	de Marsily (2009)
Units	Vol 10^3 km^3	Vol 10^3 km^3	Vol 10^3 km^3	Vol 10^3 km^3	Vol 10^3 km^3
Oceans	1.4E+06	1.3E+06	1.3E+06	1.4E+06	1.3E+06
Groundwater	6.0E+04	2.3E+04	2.3E+04	6.0E+04	1.5E+04
Of which fresh groundwater		1.05E+04	1.05E+04		
Of which brackish groundwater			1.29E+04		
Glaciers and permafrost	2.4E+04	2.4E+04	2.4E+04	3.0E+04	2.8E+04
Lakes	230	176	176	130	176
Of which fresh water lakes		91	91		
Marshes		12	11	10	
Soil water reserve	75	17	16.5	70	122
Atmospheric water	14	13	13	13	13
Biological water		1	1	2	1
Water courses	1.2	2	2	5	2
The global hydrosphere	1.5E+06	1.4E+06	1.4E+06	1.5E+06	1.4E+06
Of which fresh water reserve	5.4E+04	3.6E+04	3.6E+04	6.0E+04	3.6E+04

atmosphere in the form of vapor. While the different authors converge on the order of magnitude of flows and stocks, it will be observed that the soil water reserve is still a relatively unknown reservoir, where various estimates vary between 17,000 and 122,000 km^3.

As regards the volume of water permanently present in all of the world's rivers, it is equivalent to 2000 km^3, or 0.0002% of the water potential of the earth, an amount strictly equivalent to the volume of the Tunisian underground reserves of fresh water! But this apparent paradox is not really surprising; indeed we have seen (Table 1.1) that rivers are a very active vector of the hydrological cycle, carrying 30% of the world water balance. This represents a flux of 34,000 km^3/year, but the water flows in rivers are only in transit and remain very little time in this environment: the residence time is very short.

One can determine the water residence time in the different reservoirs of the hydrologic cycle by comparing the volume of water (Table 1.2) to the transit flow (Table 1.1). We get for the main reservoirs the following average values: (i) 2500 years for the oceans; (ii) 1500 years for groundwater; (iii) 7000 years for glaciers; (iv) 4 years for lakes; (v) 6 months for the soil water; (vi) three weeks for rivers; (vi) one week for the atmospheric water.

1.2.2 The Renewable Freshwater Resources of the Earth

The total flow, surface runoff and groundwater flow, on the continents, forms an important component of the hydrological cycle and represents the current renewable water resources. These flows maintain the balance of many ecosystems, and abstraction of freshwater resources is at the basis of much of the social and economic development of humanity. But the real renewable freshwater resources of the Earth are represented by precipitations on the continents (Fig. 1.1). Estimated at about 110,000 km^3/year on land, they perpetuate the renewable water resources of the Earth. These flows feed the continental hydrological cycle in its natural and human parts, the main term of which provides continental evaporation and evapotranspiration: through the soil water reserves, this term sustains life of continental ecosystems, and allows the development of rainfed agriculture. According to the soil moisture content and its texture, the rest of rainwater is subdivided into: (i) runoff to rivers and surface water systems, (ii) infiltration in the soil and at depth to aquifers. At the regional level, the continuous mapping of the soil water reserves is used to establish the link between meteorological and hydrological information with applications in the fields of: (i) agriculture, by detecting plant water needs for irrigation monitoring; (ii) urban hydrology and flood forecasting in case of extreme runoff conditions; (iii) hydrological and climate modeling, at regional and global scales. For all these applications, and in the same way that there are approved global networks of rainfall gauging stations, a global soil moisture database is taking place (Robock 2011).

The soil water reserve makes the link between the amount of water precipitated on land, the available energy and the carbon cycle. In terms of food production and ecosystem functioning, it is the soil water resource which, by means of solar energy and generation of photosynthetic products, determines the Terrestrial Net Primary

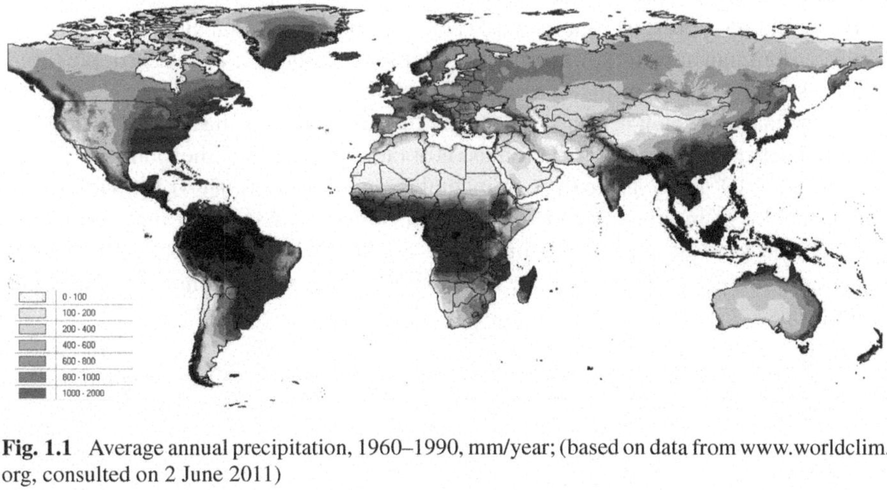

Fig. 1.1 Average annual precipitation, 1960–1990, mm/year; (based on data from www.worldclim.org, consulted on 2 June 2011)

Table 1.3 Terrestrial soils occupation and net primary production (Adapted from Michigan University 2017)

Ecosystem	Extension 10^6 km^2	TNPP dry matter 10^9 ton/year	HTNPP dry matter 10^9 ton/year	HTNPP (%)
Forests	40	48.7	13.5	28
Prairies and Rangelands	35	52.1	11.6	22
Deserts	20	3.1		
Glaciers	29	2.1		
Arable lands	16	15	15	100
Urban space	2	0.4	0.4	100
Marshes and wetlands	7	10.7		
Total continents	149	132	41	31

Production rates (TNPP), and the structure and density of the different types of vegetation on Earth. The net terrestrial primary production, a result of net energy generated by carbon sequestration on land, forms all the nutritional resources of the continents (Vitousek et al. 1986).

Global population growth, and its foreseeable development, have fostered a number of theses on the limits that land could have regarding its ability to sustainably meet nutritional needs of all its occupants. Some of these research initiatives involve the estimation of the human footprint on the biosphere and hydrosphere (Vitousek et al. 1986; Rojstaczer et al. 2001). The assessment of this impact requires estimation of the Human appropriation of Terrestrial Net Primary Production (HTNPP), especially in agriculture, urban green spaces as well as the use of wetlands, forests, grasslands, savannas, rangelands, deserts, primary production being expressed in quantity of biomass. Table 1.3 presents the estimates of the different components of the TNPP, the total of which stands at 132×10^9 tons of dry matter/year.

This table also presents published HTNPP values: most plausible estimates set the level of human appropriation of global TNPP at nearly 30% of the earth potential (Vitousek et al. 1986; Rojstaczer et al. 2001).

To translate this result in terms of water, we can estimate that the average water volume needed to produce one ton of biomass is obtained by dividing the total continental TNPP (132×10^9 ton/year) to the continental Evapotranspiration (70,400 km^3/year), namely 1.9 kg of biomass per m^3 of evapotranspiration. Assuming that this rate is constant, the level of man ownership of the continental hydrological cycle evapotranspiration term would be 31%, or 22,000 km^3/year, excluding human blue water withdrawals.

1.2.3 Water Balance and Water Withdrawals Worldwide

Worldwide we can estimate that the rainfall inland resources (110,000 km^3/year) are distributed as follows, (Ringersma et al. 2003; Molden 2007; Oki and Kanae 2006; de Marsily 2009):

(i) 57% are taken up by evapotranspiration in forests, grasslands, wetlands, rangelands, deserts, to maintain biodiversity (62,700 km^3/year)
(ii) 6% represent the evapotranspiration of rainfed agriculture as crops and livestock production (6600 km^3/year). This estimate is clarified and confirmed by the evaluation of the "green water" footprint of the World, the water equivalent of production from rainfed agriculture, calculated for 200 countries for 1996–2005 by Mekonnen and Hoekstra (2011a)
(iii) 1% corresponds to the supply of irrigated areas (1100 km^3/year), taken over by evapotranspiration.

These last three terms represent the green water withdrawals,

(iv) The remaining 36% of rainfall (39,600 km^3/year) feed the system of "blue water", including:

- 2.5% of rainfall collected for irrigation of 300 million hectares of irrigated land in the World (2750 km^3/year, half of which is actually consumed and evaporated; the rest returns back into the blue water cycle);
- 1% abstracted by cities, industries and electricity production (1100 km^3/year, 10% of which are consumed and evaporated, the rest returns to the water cycle but in an altered form).

The freshwater withdrawals

National statistics by country compiled, notably by FAO (2012), report a total freshwater withdrawals (blue water) equal to 3860 km^3/year allocated as follows: 11% for municipalities (430 km^3/year), 19% for industries (710 km^3/year), and 70% for agriculture (2720 km^3/year).

By origin, global withdrawals are distributed in the following way: 23% from groundwater (i.e. 900 km^3/year), 77% from surface water (i.e. 2960 km^3/year).

According to AQUASTAT, Groundwater supplies 38% of irrigated land in the world, but this share is highly variable; it is for example 30% in Morocco, 70% in Algeria and 77% in Tunisia.

Anthropogenic changes in the hydrological cycle

From above, the following orders of magnitude, and observations, may be concluded:

(i) Crude blue water abstractions (irrigation, drinking water, industry) amount to 3860 km^3/year, or 3.5% of rainfall and 10% of all blue water, but considering returns to the natural environment, the effective removal of blue water (by consumption) is only 1500 km^3/year, or 1.35% of rainfall and almost 4% of blue water.

(ii) Before human intervention on the water cycle, the continental inputs can be decomposed into: 64% (70,400 km^3/year) of evaporation and evapotranspiration, and 36% (39,600 km^3/year) of runoff to the oceans. Withdrawals including irrigation provoke an alteration of the hydrological cycle, by transferring a share of the runoff to evapotranspiration.

(iii) The irrigation is not the only human action having an impact on the water cycle. Other interventions also have an effect on the water cycle and in final consequence, on the sea level. Thus, the main activities to cause retention of inland waters, and therefore a lowering of the sea level, are: irrigation, through evaporation and infiltration that it induces, and the large dams reserves (30,000 facilities representing a reserve of almost 8500 km^3 worldwide). The activities that cause an increase in continental runoff, and as a result a rise in sea level, are: deforestation, withdrawal of weakly renewable groundwater and more generally groundwater overexploitation, waterproofing of urban areas and infrastructure. It is now estimated that the direct impact of human activities on the global hydrological cycle is low, resulting in a steady increase in the mean sea level whose origin is attributed to global warming, or negligible (Lombard 2007). But it is generally recognized that the effects of continental hydrology on sea level changes are uncertain, due to lack of observational data.

(iv) We can therefore assume that in a first analysis, the runoff loss due to irrigation (dams and reservoirs) is offset by the combined effects of deforestation, overexploitation of groundwater and of urbanization, so that the natural world water balance (precipitation/runoff/Evapotranspiration) remains broadly unchanged. This does not exclude the existence of an indirect impact of human activities on the global water cycle by seafloor spreading and ice melting, due to the global warming: the resulting rise in sea level is currently 3 mm/year.

The global water footprint

The latest concept of "water footprint", which has a blue water and green water components, provides a coherent and comprehensive framework for analyzing the relationship between the consumption of human societies and withdrawals on freshwater resources of the planet.

Mekonnen and Hoekstra (2011b) present a comprehensive global analysis, country by country, of the water footprint relating respectively to: (i) agricultural, industrial, and drinking water demand; (ii) agricultural use, industrial and drinking water respectively supplied locally and imported from outside. Mekonnen and Hoekstra also introduced the footprint concept of "gray water," which attempts to measure the amount of surface water and groundwater polluted by humans: the gray water footprint expresses the amount of water needed to dilute and reduce this pollution to the environmental standards.

- For all of the considered 173 countries, the total consumption represents a water equivalent of 7200 km^3/year, 6250 km^3/year of which are green water (87%), and 950 km^3/year are blue water (13%).

- 78% of the necessary virtual water in the world come from consuming countries (5600 km^3/year), and 22% come from outside (1600 km^3/year); all virtual water fluxes in the world, green and blue, are thus estimated at 1600 km^3/year.

1.3 Water Issues and Challenges in Arid Countries

The arid countries group

A land is arid when the average evaporation exceeds precipitation. In this case, vegetation suffers from lack of water (it is said to undergo water stress), flows are low and irregular, and soils degrade. The fifth of lands area-about 25 million km^2-receives less than 250 mm rainfall per year on average. These areas form what is called the "arid zone".

The International Glossary of Hydrology by UNESCO and the World Meteorological Organization (WMO) distinguish three aridity levels: (i) in semi-arid regions, rainfall is abundant enough (300 to 500 mm per year) to ensure for vegetation and runoff a seasonal rhythm; flow occurs by isolated floods, sometimes sudden and violent; (ii) arid regions receive low rainfall (100 to 250 mm per year), very irregular from year to year; intermittent runoff occurs in ephemeral streams, the wadis; (iii) in hyper arid regions, the deserts, rainfall is rare (on average 10 to 50 mm per year) and very unevenly distributed; the flow is rare, unorganized.

If we define the aridity of a country only by the criterion of the average annual rainfall, we can identify a homogeneous geo-climatic space with countries we define as arid, where rainfall is less than the threshold of the semi-arid zone with 450 mm/year. This threshold is somewhat arbitrary since immediately above are very large spaces such as Russia, Canada, where the aridity has a polar character. Figure 1.2 presents this so defined group of countries.

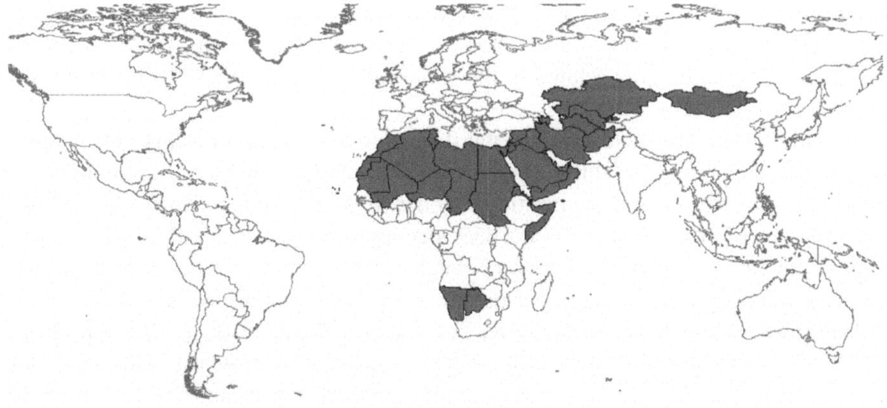

Fig. 1.2 Location of the arid countries group

Table 1.4 Water balance indicators in arid countries

	Arid countries		World		%
Mean annual rainfall mm/year	213		738		29
Rainfall resource km^3/year	5614		110,000		5
Total flow km^3/year	548	10%	39,600	36%	1.4
Evapotranspiration km^3/year	5067	90%	70,400	64%	7
Renewable internal resources km^3/year	548		43,760		1.3
Renewable external resources km^3/year	976		43,760		2.2
Blue water withdrawals km^3/year	472	86%	3850	9%	12
Total extension 1000 km^2	26,375		149,000		18
Arable land 1000 km^2	1410	5%	14,000	9%	10
Permanent cultures 1000 km^2	100	0.4%	1520	1.0%	7
Prairies and rangelands 1000 km^2	10,000	38%	33,550	23%	30
Forests 1000 km^2	1715	6.5%	40,400	27%	4.2
Non polar deserts 1000 km^2	12,000	45%	20,000	13%	60
Population millions (2007)	540		6700		8
Population millions (2050)	960		9150		10

The water balance of arid countries

It is through the AQUASTAT and FAOSTAT databases programs that we can expect to establish the water balance of the identified arid countries group. Table 1.4 brings together the key indicators related to the water balance, and allows us to locate this group in a global framework.

So while at the global level, the total flow on continents represents 36% of precipitation and evapotranspiration represents 64%, in arid countries, the total flow is limited to 10% and evapotranspiration reaches 90% of rainwater resources. Moreover, the area of arid countries is for 45% occupied by deserts (60% of the world's deserts, see Table 1.4) and 38% by rangelands (30% of the world). However, forests represent only 6.5% of the land, permanent crops 0.4% and arable land 5%.

Therefore, while they occupy 18% of the mainland, arid countries produce only 1.3% of the internal renewable water resources of the world. This indicator is to report to the population of arid countries, which represents 8% of the world population, expected to rise to 10% in 2050. Another consequence: arid countries collect 86% of their renewable water resources, which raises the question of the sustainability of this level of abstraction.

The main challenge: a difficult estimate

Reasonable management begins with the most accurate water resources assessment. For many arid regions, groundwater recharge occurs mainly from infiltration of floods in ephemeral streams: rainfall is not abundant enough to feed directly aquifers as is done in temperate climate and infiltration is only possible when runoff concentrates as floods in the beds of wadis. Rainfall and runoff are greatly variable depending on the year: an unusually humid episode, causing the aquifer reserve replenishment, may be followed by a dry period of several decades, during which groundwater recharge drops significantly. Therefore, assessment of surface and groundwater resources requires availability of long series of precipitations, runoff and groundwater levels: the longer the series, the accurate is the water resource assessment. The great rainfall variability refers to the need for saving water volumes in order to defer its use during drought.

How do we evaluate the aquifer resources? In temperate climates, the resource estimate is mainly done by measuring superficial flows, observable at the aquifer outlets: springs and mainly rivers. In arid areas, the aquifers outlets are diffuse, consisting of large evaporative systems-chotts and sebkhas-whose flows are very difficult to measure.

For water needed to the development of arid regions, estimates of hydrogeologists are generally followed by immediate production. Given the lack of sufficiently long measurement series, the estimation of exploitable quantities progresses like the quantities actually being withdrawn: hydrogeologist opts for the precautionary principle and is considered as the champion of aquifers under-exploitation…But as soon as this resistance goes down, overuse appears.

Moreover, the arid regions contain large sedimentary basins conducive to the accumulation of sometimes considerable groundwater reserves: one can cite, for example in Africa, the North-Western Sahara Aquifer System, the Nubian Sandstones Basin, the basin of Taoudeni and that of Lullemenden (see Fig. 1.3). The evaluation of exploitable resources is even more difficult in hyper-arid regions. Saharan larger basins, for example, contain enormous freshwater reserves, estimated at more than 400,000 billion cubic meters, but only part of them are sustainably exploitable due to their low recharge rate. How far these deep reserves can be withdrawn?

During the last decades, the economic development needs have boosted a mining and economist vision of these aquifers exploitation and this approach found a wide application field with the evolution of numerical models prediction capabilities. This helped to validate important decisions concerning mining aquifer exploitation on the basis of a rudimentary recharge knowledge (North-Western Sahara aquifer) or neglecting present recharge (Nubian Sandstones aquifer), agreeing that water levels in wells undergo continual drawdown over decades, even centuries.

Compared to the gigantic reserves volumes, knowledge of large deep aquifers recharge has become accessory: for fifty years, the separate and parallel develop-

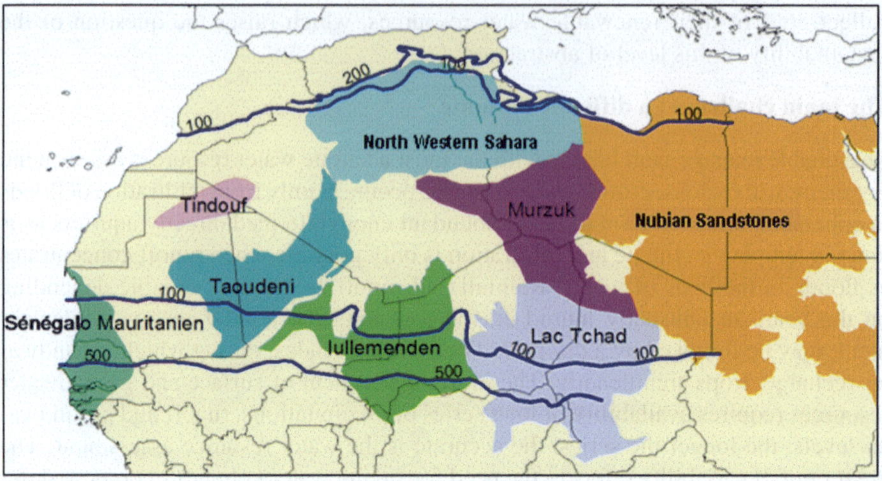

Fig. 1.3 The Saharan Aquifer Basins and average precipitations, mm/year (author's elaboration, data from OSS and Margat 1995)

ment of surface and groundwater hydrologies, the deep waters age determination-estimated under certain conditions to tens of thousands years-and the difficulty of an experimental approach by in situ recharge observation helped to establish the image of water mines under deserts, and strengthened the concept of fossil water-water masses accumulated in earlier wetter periods without any present recharge. However, a significant current recharge of these aquifers has been demonstrated through recent works (BabaSy 2005; Gonçalvès et al. 2013). Water management also involves assessing the risks of overexploitation, water quality degradation, and extreme events: floods and droughts, large dams construction, etc.

Drought prevention and management

Drought is a climatic event which results from an unusual situation of atmospheric circulation, leading to a precipitation deficit. This situation has a direct effect on the vegetation, which characterizes the agricultural or edaphic drought (in relation to the soil water reserve). When the deficit occurs in a very intense or extensive manner, the recharge of surface water and ground water components within a catchment area is affected, resulting in a deficit in water flow in water courses or a lowering of the groundwater level: it is the hydrological drought. Furthermore, distinction should be made between the predictable structural droughts which affect a more or less important part of a region, and the exceptional droughts, which concern all the natural and cultivated vegetation and most of the territory of a given country.

Determining the impacts of droughts requires monitoring economic, social and environmental indicators. The monitoring of hydro climatic variables facilitates the announcement of drought and the initiation of adequate responses (Bargaoui 2002); it is essential for drawing up balance sheets whose cartographic presentation constitutes

Fig. 1.4 The flood of Oued Zeroud at Sidi Saad, Tunisia, September 1969 (from Kallel et al. 1972)

one of decision-making aid elements. In the USA, the National Oceanic and Atmospheric Administration (NOAA) regularly supplies the National Integrated Drought Information System (NIDIS[2]) with climate and hydrological data, which are available to the public. The NIDIS is itself managed by the National Drought Mitigation Center which disseminates a set of maps showing the state of the drought indicators as well as the forecast bulletins elaborated by the Center.

Drought management has two dimensions: crisis management and long-term management. The setting up of observatories with monitoring and assessment functions for the environmental, economic and social impact of droughts is a basic tool for stakeholders. FAO's Global Information and Early Warning System, FAO's GIEWS,[3] are currently effective means of predicting and assessing drought in the international community.

Exceptional floods and surpluses management

Floods affect all regions of the world, but they are particularly violent and destructive in the arid zone. For example, the Oued Zeroud flooding, in central Tunisia, on 27 September 1969, had a maximum flow of 17,000 cubic meters per second (Fig. 1.4), seven times higher than the historic flood of 1910 in Paris, (2400 cubic meters per second) for a five times smaller watershed.

---

[2]https://www.drought.gov/drought/what-nidis, consulted 2017, Jan 15.

[3]http://www.fao.org/es/giews/english/index.htm, consulted 2017, Jan 15.

The exceptional floods of 1969 had a return period estimated at 150 years. In general, however, the variability of surface water inputs is very high: for the country as a whole, the surface water runoff associated with the centennial recurrences vary from 637 million m^3 in dry periods to 8510 million m^3 in wet periods (Frigui and Touzi 2009). Although hydraulic infrastructure is designed to allow inter-annual management of water resources, which provides a form of "water saving for drought", they do not, on the other hand, allow for the collection of water surpluses that flow in wet periods. Indeed, one year out of 10, the volumes of surface water amount to 4760 million m^3, a surplus of about 2000 million m^3 compared to the estimated average value, 2750 million m^3 (Frigui and Touzi 2009). Given the capacity of water infrastructure, these surpluses are not controlled and this resource flows to the sea or to natural outlets such as sebkhas. These figures should challenge the actors of the water sector on the desirability of controlling these inputs and taking them into account in water balances, in particular to improve the water budget of over-exploited aquifers.

Developing new water resources in arid countries

The renewable resource mobilization rate has become higher in many arid countries: 85% in Tunisia, 65% in Morocco, 60% in Algeria. To increase the water potential of such countries, there is a need to improve the water cycle efficiency, or to develop non conventional processes in order to exploit other resources. Building hydraulic facilities remains an important strategy, but is now considered at a smaller scale. Water mobilization by large dams is on the way to completion in all North African countries. Large hydraulic facilities are gradually complemented by a new generation of small-scale works, inspired by traditional facilities, and can be managed by users: hill lakes, that is to say, small water reservoirs as earth dams, watershed erosion control by banquets, cisterns. These traditional techniques for water recovery are again promoted in arid regions, although they remain vulnerable to drought, and their actual impact on the hydrological cycle (aquifer recharge, feeding dam reservoirs) is not always well known.

Another route is the long-term storage in aquifers, experienced for many years with good results by the arid states of the US, including the Arizona Water Bank Authority. It is true that they have taken advantage of this unique source that is the Colorado River. Elsewhere in the arid world, water sources of that size are rare, and underground storage experiments are still modest. One can cite experiences on wadi beds in Tunisia and Oman, but the most complete site is probably that of the Souss valley in Morocco, where infiltrated quantities in the wadi beds, by reservoir releases from the Aoulouz dam in the High Atlas, totaled one billion cubic meters in 12 years.

The development of alternative resources is another way of increasing supply. Two techniques are used: desalination and wastewater reuse. The North Africa and Middle East region has now 60% of the world seawater and brackish water desalination capacity, half of which in Saudi Arabia. This region produces 2.5 billion cubic meters of water per year. This has contributed to lower costs, thanks to scale economies due to the facilities size, and the increasing efficiency of energy recovery systems, which represent over 50% of total costs. Becoming competitive, desalination is adopted for drinking water supply of the coastal cities, in the oil countries, but also in other arid countries.

Finally, the treated wastewater can fill a variety of uses, whether agricultural, recreational, industrial or environmental, for example to recharge groundwater, or maintain wetlands. Experience of arid countries in treated wastewater reuse for agricultural purposes is old, but treated wastewater reuse remains modest because of the difficulties inherent to its operating mode: these waters which are indeed expected to pose a health risk, are often poorly perceived by users, and produced far from centers of agricultural use. Despite these challenges, the reuse of treated wastewater is required as an essential alternative. In 2006, WHO, in cooperation with FAO and UNEP, published the third edition of the "Recommendations for the Reliable Use of Wastewater in Agriculture and Aquaculture". However, consumers' acceptance and perception of agricultural products grown with treated wastewater generally remain negative.

1.4 The Impacts of Anthropogenic Activities

The risks to water resources can be caused by natural hazards (droughts, floods), they also result from anthropogenic action, often in relation to the management and use patterns of the water and soil resources. The impacts of anthropogenic activities on water resources are particularly critical in arid or semi-arid countries where strategies for mobilizing and massive water transfers exacerbate the problems of water scarcity: increased mobilization of water resources, water quality deterioration, overexploitation of underground aquifers, irrigation impacts on salinization of soils and groundwater, impacts of water dams reservoirs on groundwater recharge.

Degradation of water quality

Water quality refers to the set of physical, chemical and biological characteristics required for a given water use. Pollution modifies the physical, chemical, and microbiological characteristics of water, reducing its uses. In Europe, agriculture is now responsible for a large part of pollution. While the control of point source pollution (industry, collectivities) has been effective due to significant investments in wastewater treatment, the large-scale use of fertilizers and pesticides in agriculture creates a diffuse pollution that may cause eutrophication of water environments. The situation in developing countries is not comparable: pollution of urban water has adverse effects on health and the environment, often due to the rudimentary state of clean water and sanitation facilities.

Water quality protection has been recognized at the international level and the general principles of water resources management systematically integrate qualitative aspects. In particular, chapter 18 of Agenda 21 states that the general objective of protecting freshwater resources and their quality is "to make certain that adequate supplies of water of good quality are maintained for the entire population of this planet, while preserving the hydrological, biological and chemical functions of ecosystems, adapting human activities within the capacity limits of nature and combating vectors of water-related diseases". At the European level, the Water Framework Directive (Directive 2000/60/EC) defines water management activities in a global vision of water environments that combines ecological, hydrological, socio-economic, technological and institutional aspects. The "water good status" qualitative and quantitative criteria are defined by considering the aquatic environment state of reference and setting objectives with achievements deadlines.

Aquifers overdraft

Aquifers mitigate climate variability by their storage capacity and constant availability, and enhance the water supply of many users by their distribution in space, often at shallow depths. But if the local impact of individual withdrawals is limited, their indiscriminate multiplication leads to overexploitation. Groundwater overexploitation refers to a state where, over a multi-year period, the abstraction from a given aquifer exceeds the recharge flux. When this imbalance persists for several decades, the exploitation regime is unsustainable, resulting in depletion of reserves and exhaustion of the aquifer. Depletion of aquifer reserves (Fig. 1.5) has been observed in western and central USA, India, Pakistan, Iran, Northern China, Saudi Arabia, North Africa and Southern Spain. However, both water abstraction and aquifer recharge are difficult to evaluate accurately, particularly in arid and semi-arid zones: when the exploitation reaches orders of magnitude close to the recharge, the excess of water abstraction is then taken in the margins of uncertainty, and only long and detailed observations and analyzes of the aquifer behavior make it possible to specify the diagnosis of overexploitation.

Worldwide groundwater withdrawals are estimated at 1000 km^3/year in 2010 (Van der Gun 2012), of which 67% for irrigation, 22% for domestic needs and 11% for industry. These volumes account for almost 10% of renewable groundwater resources and 25% of the world's freshwater withdrawals. Figure 1.5 shows the evolution of abstractions in the 8 strongest abstractors, whose sum in 2010 was more than 66% of the world total, with near 60% for the Asian continent alone: India takes 250 km^3/year, by nearly 25 million wells and boreholes. The volume of water supplied by overexploitation of the world's aquifers, which is of recent interest due to its proven impact on sea level rise after the slowdown in the construction of large dams (Wada et al. 2016), is estimated between 110 and 200 km^3/year (Konikow 2011; Wada et al. 2012; Döll et al. 2014). Although still uncertain, overexploitation of aquifers is expected to continue in the 21st Century due to climate change and rising temperatures, resulting in increased irrigation requirements.

Many shallow aquifers in arid regions were exploited unduly to support the development of irrigated agriculture, in certain parts of India for example. This resulted in significant drawdowns in some aquifers, sometimes irreversible, often accompanied with a slow chemical quality degradation by salts leaching from irrigation or from rocks, or by sea water intrusion in coastal aquifers. This strikes domestic assets of

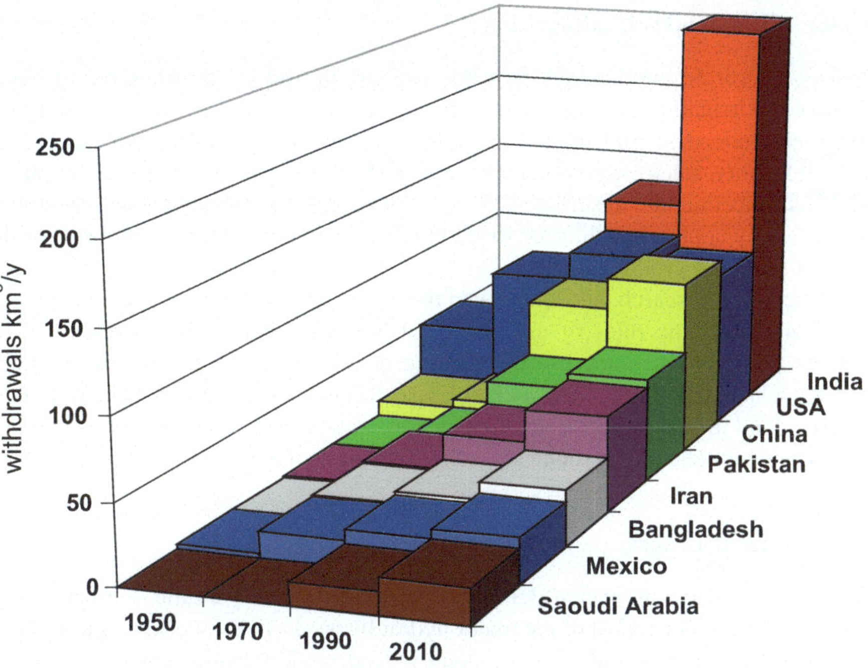

Fig. 1.5 Groundwater withdrawals (km^3/year) in countries with highest abstractions from 1950 to 2010 (Authors' elaboration, data from: Shah 2004; Margat 2008; Van der Gun 2012)

arid countries: the economic activities developed by this resource certainly contribute to GDP growth in the short term, but the water resources overuse weakens national heritage and reduces the natural wealth of the country. This highlights the "common good" concept limitations: each user finds his individual interest to take ever more on a shared resource. To prevent abuse, negotiation processes between users should be developed, to limit the groundwater degradation.

In order to resolve conflicts of use and avoid overexploitation, either of these two approaches is generally proposed: either through the establishment of a water market with prices that are imposed on all, or to resort to State intervention to ensure an equitable distribution of resources. The reality is that the effectiveness of water markets to contain overexploitation has not been demonstrated (Petit 2004). As regards the regulatory instruments implemented or supported by the State, they are of different types: regulatory (authorizations, prohibitions), economic (royalties), participatory (users participation, supervised by public authority). In order to contain the overexploitation of natural resources, governance processes based on negotiation, participation and consultation between users are developed (Ostrom 1990). But on aquifers in the arid zone and with the exception of some local particular initiatives, the acuity of the problem and the recent awareness of the gravity of the phenomenon have not yet led to the emergence of effective and sustainable remediation strategies.

Water salinity and soil salinization

Besides overdraft, another risk is water salinity. In arid countries, some of water allocated to irrigation is naturally rich in salts and considered as poor regarding the crop yield. Indeed, salinity may decrease the yield or quality of crops and may reduce the soil fertility, since evapotranspiration gives back pure water into the atmosphere, while salts accumulate in surface soil and root zone. These salts excesses must then be washed by applying additional irrigation doses and good drainage, but part of the leached salts join groundwater.

Agricultural research has determined the conditions and limits of brackish water exploitation, but the rules of use and good practice are not always respected by farmers: there is still underuse of leaching doses, poor maintenance of drainage networks, lack of drainage and water table rise. It is now estimated that a quarter of the irrigated land in arid countries are damaged by water salinity, which is considerable. User's behavior change is crucial regarding the future of irrigated agriculture in these regions.

Large water transfers

Transfers constitute an ancient fact, peculiar to the great hydraulic civilizations of the arid regions. The oldest of the major hydraulic works in the world was to transfer virtual water: on the great Chinese canal indeed, undertaken as early as 700 BC. JC and over 1800 km in length, thousands of ships circulated every year to supply Beijing with cereals from the southern provinces. Modern water transfers by canals or pipes, taking surface or underground waters for irrigation or drinking water, are no less impressive in terms of distances traveled and flows transported: the Karakoum Canal in Turkmenistan (1300 km and 11 km^3/year), Great Artificial River of Libya (2000 km and 3 km^3/year), Feeding of Tamanrasset (700 km and 60 million m^3/year). The large internal transfers in the USA represents a total flow rate of 850 m^3/s and those of Canada 4400 m^3/s, the latter being designed for hydroelectric generation (Lasserre 2005). In Tunisia, the 160 km long Medjerda-Cap-Bon canal feeds large irrigated perimeters in Cap-bon, and supplies drinking water to major coastal cities.

But the biggest project is undoubtedly the South-North Transfer in China, part of the waters of the Yangtze River is transferred to the less resource-rich northern rivers, where the Yellow River suffers from chronic water deficit. The hydraulic scheme, which has just started after fifty years of study and which required an investment of 45 billion Euros, includes three parallel drifts of more than 1000 km each, the eastern branch running along the Grand Canal, carries a flow of 45 km^3/year.

The extent of water transfers is reflected in the investment costs and impacts in terms of the socio-economic development of the regions, but also in the hydrological and ecological issues they can create: watercourses and Wetlands drying up, ecosystem disruption. Among the most notorious changes which caused environmental disasters one can cite those of the Colorado River in the US or the Syr and Amou Daria in Central Asia, which resulted in the destruction of the delta marshes of Colorado and the disappearance of the Aral Sea. On the other hand, the lack of equity and the sometimes unbalanced socio-economic development between the surplus and deficit regions concerned by water transfers, may constitute a source of virulent opposition.

Impact of reservoir dams on groundwater

The sudden and devastating floods, alternating with periods of prolonged drought, have prompted many arid countries to secure their supply by building large dam reservoirs. However, these large facilities are not free of risks, such as silting and drying up natural recharge sources of downstream aquifers. Thus, the dams in the Saharan Atlas built upstream of the North Western Sahara Aquifer System have a direct impact on the aquifer recharging mode. In principle, the dams can supply the aquifer recharge function, by reservoir releases, that is to say by artificial and controlled evacuations of retained water, which will infiltrate into the wadi bed for aquifer recharge. However, this function is rarely assigned to the dams, whose main function is irrigation water supply, a binding function that requires a great supply security. The operating strategies of these facilities leave no opportunity to recharge in wadis, except in exceptional years, when the dam is brought to dump.

Already when recharge function is identified and assigned at the design of the dam, but is shared with irrigation, the experience of Kairouan in Tunisia (Nazoumou 2002) shows that due to the water supply requirements for farmers it is difficult to guarantee recharge, relegated to an accessory level.

Climate change

Climate change refers to a long-term change in average global weather attributed directly or indirectly to human activity that alters the composition of the global atmosphere. Climate change shall be added to the natural variability observed in comparable periods. The Mediterranean basin is among the most vulnerable regions to the impacts of climate change. Agricultural activity could prove to be one of the most threatened economic sectors. The countries of the southern and eastern shores are likely to be the most severely affected, resulting in an increase in food dependence.

Water is the vector through which the effects of climate change are felt by people, ecosystems and economies. In its various reports, the Intergovernmental Panel on Climate Change (IPCC 2008, 2009; Collins et al. 2013) highlights the significant impacts of this change on the state of water resources with changes in quality, vulnerability of ecosystems, increased water demands. These effects will add to the already complex challenges of water management in arid and semi-arid regions. In fact, responses to climate change, in the form of mitigation or adaptation programs, are an integral part of decision-making on the sustainable management of water resources and should be integrated into national economic development planning, social and regional development. The climate change adaptation strategy presents an additional challenge for the poorest countries.

1.5 Principles, Concepts and Instruments of Water Policies

This section presents the main principles, concepts and instruments of water policies. Among the basic principles on which water policies in some countries are based, it should be mentioned the principle of "user pays" and "polluter pays", the precautionary principle and the principle of subsidiarity. As regards the instruments of water policies, and among the most important, the different management approaches are presented: Integrated Water Resources Management (IWRM), water demand management, watershed-based water management, participatory management. Mention is also made of aspects relating to the economic dimension of water (value and pricing) as well as water data and information systems on water

1.5.1 Principles and Concepts

The "user pays" and "polluter pays" principles

As with any common natural resource, the exploitation of water resources generates socio-economic benefits but generates a real cost for society, related to the quantitative and qualitative degradation of the resource (costs of drinking water treatment, Wastewater treatment, etc.). Economists talk about external costs or externality. It is currently recognized that any policy to protect the environment must involve internalization of environmental costs. This theory, developed by the economist Arthur Pigou, forms the basis of the "polluter pays" principle.

This principle covers several categories of payers (user-payer, collector-payer, consumer-payer, etc.). The legal formulation of the polluter pays principle was developed by the OECD in 1972. At the international level, the Rio Declaration (principle 16) states: "National authorities should endeavor to promote the internalization of environmental costs and the use of economic instruments, taking into account the approach that the polluter should, in principle, bear the cost of pollution, with due regard to the public interest and without distorting international trade and investment".

The analyses and recommendations of the Organization for Economic Co-operation and Development (OECD 1992) stressed the difficulties of implementing the polluter-pays principle; In particular with regard to the definition of the polluter: who in the economic chain (from the supplier to the consumer via the producer and the distributor) must be considered as polluter and how to decide between the responsibilities of each? On the other hand, while it is easy to identify pollution and qualify it for localized pollution, this task becomes difficult when it comes for diffuse sources, such as pollution from agricultural uses.

The user-pays principle aims at better involving users by transferring the costs of using water resources to consumers in order to reduce public expenditure. The principle of "user-pays", based on water pricing and water taking fees and charges, makes it possible to improve the efficiency of water use and to ensure the conservation of the resource by controlling water withdrawals.

The precautionary principle

The precautionary principle has existed for a long time in the field of public health. It emerged in the field of the environment at the end of the 1960s in Federal Germany, where scientists and policy-makers tackled the problem of forest decline with the recommendation to the public authorities to take all "necessary and reasonable measures" to address potential risks even without adequate scientific knowledge, (Harremoës et al. 2001). This principle was formulated in the Rio Declaration in 1992 and introduced into Community law by the Treaty of the European Union which specified the framework for environmental policy "based on the principles of precaution and preventive action against environmental damage...". The precautionary principle is considered as an integral part of the concept of sustainable development. It consists in putting in place measures capable of preventing (at economically acceptable cost) the risks of serious and irreversible damage, where scientific and technical knowledge does not provide certainty, mainly in the fields of environment and health. Unlike prevention that focuses on proven risks, precaution, a form of caution in action, focuses on potential risks (climate change, industrial pollution, contamination of groundwater, cultivation of GMOs, etc.).

The principle of subsidiarity

The principle of subsidiarity aims to ensure that decisions are taken as close as possible to the people they concern, in a process that involves all the actors concerned, favors consultation and coordinates means and actions. Its application leads to the decentralization of the authority: the State delegates some of its powers when it considers that they can be better assumed on local scales in view of their proximity to the citizens. In the field of water, its application induces a decision at the level closest to the places of use or degradation of the resource. For example, water management by hydrological basin or aquifer basin can be considered as subsidiary: social actors define and realize the general interest via the basin committee. This system is at the origin of the construction of a collective responsibility. It brings together users of a common heritage who are thus bound by natural solidarity (Barraqué 1997).

1.5.2 Water Policy Instruments

Integrated water resources management

The term "Integrated Water Resources Management" (IWRM) emerged in the aftermath of the international conferences on water and the environment in Dublin and Rio in 1992, but this concept had not been clearly defined. Thus, in 1996, the World Water Council (WWC) and the Global Water Partnership (GWP), were created almost simultaneously, respectively in Marseilles and Stockholm, to precisely promote the paradigm of Integrated Water Resources Management (IWRM).

GWP defines IWRM as "a process which promotes the coordinated development and management of water, land and related resources, in order to maximize the resultant economic and social welfare in an equitable manner without compromising the sustainability of vital ecosystems". It is sometimes described more simply as a process aimed at improving the economic efficiency of water use, promoting social equity in access to water and ensuring environmental sustainability.

Adopted by all as the only framework for the design and evaluation of water policies, IWRM has established itself as the unavoidable model of governance, so that there is no longer any water project or policy which does not claim it. This propensity to universalism, however, is limited by some considerations:

(i) The perfection of this conceptual model has become such that, in practice, it diminishes the chances of eligibility, if not to deploy means that exceed the capacities of many countries, particularly developing countries. Thus, IWRM is rather a sum of objectives to be attained. Its adoption as a mode of comprehensive management mode has been at the origin of institutional transformations: only a few countries, mostly of the OECD, have radically restructured their legal structures and institutional frameworks in this direction (Petit 2009). Elsewhere, intermediate objectives are set by the formulation of integrated management action plans or PAGIRE: institutional reforms to create an enabling environment for IWRM. (ii) There are as yet no clear indicators of whether a system is in compliance with IWRM or is in a transition state; research is underway in this field, (UN-Water 2010).

Water demand management

Adjusting the water supply to the water demand has long been the only way of managing the resource. "Supply management" traditionally deals with development and exploitation of additional water resources (conventional or non conventional). "Water Demand Management" focuses on mechanisms to achieve more effective levels and modalities of water use. On a practical level, it covers a wide range of instruments classified into four main categories: (i) Technical instruments with objectives for improving performance of hydraulic systems: reduction of water losses in transport and distribution systems, improvement of quality of service, rationalization of water use at the users level, promotion of efficient technologies, water recycling, (ii) Economic instruments such as pricing, cost recovery, financial incentives to guide consumers, public-private partnerships in water management, market mechanisms; (iii) Legislative and institutional instruments for the protection of water resources, the quantitative restrictions and water rationing, the decentralization of water management, the regulation and coordination of the public services, the participation of the users in the management; (iv) Measures aiming to change the users behavior with regard to the water resources conservation and use through information, awareness-raising and stakeholder mobilization programs, and capacity building (Brooks et al. 2007).

Water demand management has become an unavoidable necessity in countries where the mobilization of water resources is capped, while the demand is increasing. It aims to rationalize and optimize use in all sectors. In this context, the role of the normative system is essential: national standards for water use, promotion of appropriate technologies, and implementation of conformity assessment activities. According to ISO (International Organization of Standardization), the standard is "is a document that provides requirements, specifications, guidelines or characteristics that can be used consistently to ensure that materials, products, processes and services are fit for their purpose". The standardization of hydraulic equipments and accessories contributes to the valorization of water and to the fight against water wastage. It provides stakeholder with the necessary technical and economic information to improve the efficiency of hydraulic systems. Strengthening standardization (including conformity assessment activities) and promoting their application can provide valuable tools for achieving the objectives of water use control and optimization.

Watershed-based water management

Watershed-based water management was introduced for the first time in France by the 1964 law that created the Basin Agencies. Recommended in various international conferences (Dublin, Rio, Kyoto), this mode of management has been adopted by many countries. The 2000 European Water Framework Directive goes even further and considers the river basin as an ecosystem that covers hydrological, ecological, socio-economic, technological and institutional aspects. But the implementation of such a model is a challenge in many countries where administrative organization and political design are not conducive to decentralization and effective user participation. Water laws in Morocco and Algeria have introduced water management within watersheds and created basin agencies. Nevertheless, the management modes remain largely under the control of the central administration, which defines the strategic choices for the mobilization, allocation and use of resources. In Tunisia, various studies related to the water sector have referred to the principle of integrating water management within entities respecting the hydrological cycle, but the centralized mode of water management and the high degree of artificialization of the hydrological cycle have not yet provide concrete and applicable responses to the question of the geographical framework of water management.

Participatory management

Despite its complex nature, social participation in the field of water management and governance has become a reality in many countries and can take many forms in practice. One can try to characterize it in two significant situations: (i) public participation in river basin management planning: for example, European citizens are called upon to play a key role in the implementation of the Framework Directive On Water, which invites public information and participation in the development of river basin management plans to improve water quality. This is one of the rights provided for in the Aarhus Convention, which codifies access to information, public participation in decision-making and access to justice in environmental matters; (ii) water user associations: in rural irrigation or drinking water projects, user participation is a factor of efficiency and sustainability. The success of user associations depends on the democratic context in which these associations operate, as well as on the government support for technical assistance, capacity building and conflict resolution.

The economic dimension of water: value and pricing

The economic dimension of water has two elements necessary for rational resource planning: (i) The economic value of water for different categories of uses: the value of water for domestic use is often determined by the "Consent of consumers to pay". For industrial and agricultural uses, the value of water is estimated indirectly through the economic value of the commodities produced; (ii) The cost price of water at the point of delivery: it can cover operating and maintenance costs (water supply, wastewater treatment) as well as investments in infrastructure.

Despite the various reservations (religious, economic, moral), it is currently accepted that water has a cost, and if it is not the user who pays it, the costs are incumbent upon the State, the decision being in reality of a political nature. Given this principle, the aim of any tariff regime remains the financial self-sufficiency of the body responsible for the management of water systems, which must have sufficient financial resources to ensure the sustainability of investments. This does not preclude certain targeted consumer groups from being subsidized in a transparent way by others for social, strategic or logistical reasons. Neither can it be excluded that certain considerations related to other objectives such as incentives to the preservation of the resource, rationalization of the use, for being introduced to better guide the behavior of the consumers, (Dinar and Subramanian 1997).

Water data and information systems

Improving the decision-making capacities of operators in the water sector, through better observation and more efficient resources management, results in the establishment of a national water information system. Through all the data it generates and transforms into information that can be used by decision-makers, government and users, this information system is central to the process of formulating national water policies. These are based on the development of structured databases, dynamic and interactive geographic information systems, and powerful modeling tools. The modernization of the national water information system facilitates the widest public access to information and provides reliable data for a better assessment of water policies.

To be efficient, this modernization of information systems is based on adequate legislative and institutional instruments. Thus, the most recent laws have fully integrated the need to base the development of national policies on the operation of information systems. In Europe, the WISE: Water Information System for Europe, developed by the European Environment Agency, maintains and disseminates data and information on water quality, resources and uses in European countries. At the national level, countries are also equipped with accessible water information systems to promote transparent water and environmental policies. In the United States, the USGS is in charge of developing water information: it records the occurrence, quantity, quality, distribution, movement and use of surface water and underground and disseminates data to the public, administrations, authorities and agencies involved in water resources management.

National water information systems are now aligned with the new strategic challenges of water policies, where information plays a major role in: (i) assessing the state of hydrological environments and pressures; (ii) limiting and preventing the consequences of hazards and risks (droughts, floods, degradation of resources, over-exploitation).

References

BabaSy M (2005) Recharge et paléorecharge du système aquifère du Sahara septentrional. Ph.D. Thesis, University of Tunis El Manar, Tunis

Bahri A (2012) Integrated urban water management. Global Water Partnership Technical Committee. Background papers

Bargaoui Z (2002) Prévention et gestion des sécheresses. In: Besbes M et al. (2002) Water future, a new challenge for Tunisia, Int. rep. Tunisian institute for strategic studies, Tunis

Barraqué B (1997) Subsidiarité et politique de l'eau, in Faure: Territoires et subsidiarité, l'action publique locale à la lumière d'un principe controversé. L'Harmattan, Paris

Besbes M (2012) L'humanité va-t-elle manquer d'eau? Le Figaro, Paris, 19 Janvier 2012

Brooks D, El-Fattal D, Arafa D (2007) Gestion de la demande en eau, définitions et critères pour évaluer l'efficacité de la GDE et exploration des notions d'innovation sociale et d'économie politique, WADMmena–IDRC

Chahine M (1992) The hydrological cycle and its influence on climate. Nat 359(6394):373–380

Collins M, Knutti R, Arblaster J, Dufresne J-L, Fichefet T, Friedlingstein P, Gao X, Gutowski WJ, Johns T, Krinner G, Shongwe M, Tebaldi C, Weaver AJ, Wehner M (2013) Long-term climate change: projections, commitments and irreversibility. In: Working group i contribution to the IPCC fifth assessment report (AR5), climate change 2013: the physical science basis

de Marsily G (2009) L'eau, un trésor en partage, Dunod, 2009, Paris

Dinar A, Subramanian A (1997) Water pricing experiences. An international perspective. World bank technical paper, World Bank

Döll P, Muller Schmied H, Schuh C, Portmann FT, Eicker A (2014) Global-scale assessment of groundwater depletion and related groundwater abstractions: combining hydrological modelling with information from well observations and GRACE satellites. Water Resources Research 50:5698–5720. https://doi.org/10.1002/2014WR015595

FAO (2012) AQUASTAT: FAO's global water information system, the land and water division. http://www.fao.org/nr/water/aquastat/main/index.stm

Frigui L, Touzi S (2009) Caractérisation des événements extrêmes en Tunisie et possibilités de la maîtrise des excédents pour une gestion durable de la ressource. in: l'Etat des ressources en eau du Maghreb en 2009. UNESCO, Unesco office in Rabat

Gleick PH (1996) Water resources. In: Schneider SH (ed) Encyclopedia of climate and weather, vol 2. Oxford University Press, pp 817–823

Gonçalvès J, Petersen J, Deschamps P, Hamelin B, Baba-Sy O (2013) Quantifying the modern recharge of the "fossil" Sahara aquifers. Geophys Res Lett 40:2673–2678. https://doi.org/10.1002/grl.50478

Harremoës P, Gee D, Macgravin M, Stirling A, Keys J, Wynne B, Guedes Vaz S (2001) Late lessons from early warnings: the precautionary principle 1896–2000. In: Environmental issue report no. 22. Copenhague: Agence européenne pour l'environnement

Harvey C (2000) Groundwater Hydrology. Water balance equation; http://ocw.mit.edu/courses/civil-and-environmental-engineering/1-72-groundwater-hydrology-fall-2005/lecture-notes/1_72_lecture_1.pdf
IPCC (2008) Climate change and water, technical report VI. Inter governmental Panel on Climate Change
IPCC (2009) The copenhagen diagnosis. Updating the world on the latest climate science. Interim Report Inter governmental Panel on Climate Change, Climate Change Research Centre
Kallel R, Colombani J, Eoche-Duval JM (1972) Les précipitations et les crues exceptionnelles de l'automne 1969 en Tunisie. Ressources en Eau de Tunisie 2:48–120
Konikow LF (2011) Contribution of global groundwater depletion since 1900 to sea level rise. Geophys Res Lett 38:L17401. https://doi.org/10.1029/2011GL048604
Korzoun VI, Sokolov AA (1978) Water management and development. In: Proceedings of the United Nations Water conference. Mar del Plata, Argentina March 1977, vol 1, part 4. Pergamon Press, New York, p 2199–2215
Lasserre F (2005) Les projets de transferts massifs d'eau en Amérique du Nord—Vertigo, Hors-série 2/Septembre 2005
Lombard A (2007) Les variations actuelles du niveau de la mer, observations et causes climatiques. La Météorologie, N. 59, Nov 2007
L'vovič MI (1968) Le bilan hydrique du globe terrestre. Annales de Géographie 77(423):553–566
Margat J (2008) Exploitation et utilisation des eaux souterraines dans le monde, Ed BRGM, France
Mekonnen MM, Hoekstra AY (2011a) The green, blue and grey water footprint of crops and derived crop products. Hydrol Earth Syst Sci 15:1577–1600
Mekonnen MM, Hoekstra AY (2011b) National water footprint accounts: the green, blue and grey water footprint of production and consumption. In: Value of water research report series no. 50. UNESCO-IHE, Delft, The Netherlands. http://www.waterfootprint.org/Reports/Report50-NationalWaterFootprints-Vol1.pdf
Michigan University (2017) Global change—science of sustainability. http://www.globalchange.umich.edu/globalchange1/current/lectures/kling/energyflow/
Molden D (2007) Water for food, water for life: a comprehensive assessment of water management in agriculture. Earthscan/International Water Management Institute, London
Nazoumou Y (2002) Impacts des barrages sur la recharge des nappes en zone aride. thèse de Doctorat, Ecole Nationale d'Ingenieurs de Tunis, Tunis-El Manar University, Tunisia
OECD (1992) The polluter-pays principle, analyses and recommendations, vol 92, p 81. OECD/GD, Paris
Oki T, Kanae S (2006) Global hydrological cycles and world water resources. Sci 313(5790):1068–1072
OSS, Margat J (1995) Ressources en eau communes des pays de la région de l'OSS—Bassins fluviaux et aquifères profonds transfrontaliers, Observatoire du Sahara et du Sahel, Tunis
Ostrom E (1990) Governing the commons. The evolution of institutions for collective action. Cambridge University Press, Cambridge
Petit O (2004) La surexploitation des eaux souterraines, enjeux et gouvernance. Nat Sci Soc 12:146–156
Petit O (2009) La politique de gestion des eaux souterraines en France. Une analyse en termes de gestion intégrée. Économie rurale 2009/1, no. 309
Ringersma J, Batjes N, Dent D (2003) Green water: definitions and data for assessment. Report 2003/3. Wageningen: ISRIC world soil information. http://www.isric.org/isric/webdocs/Docs/ISRICGreenwater%20ReviewFebr2004_.pdf
Robock A (2011) A new international network for in situ soil moisture data. Eos 92:17, 26 April 2011
Rojstaczer S, Sterling S, Moore N (2001) Human photosynthesis products. Sci 294:2549
Shah T (2004) Groundwater and human development: challenges and opportunities in livelihoods and environment. Presentation, made at the 2004 Stockholm world water week. Water Sci Technol

Shiklomanov IA (1993) World fresh water resources. In: Gleick PH (ed) Water in crisis. Oxford University Press, Oxford, pp 13–24
Shiklomanov IA (1998) World water resources: a new appraisal and assessment for the 21th Century. International hydrological programme, UNESCO
Shiklomanov IA, Sokolov A (1983) Methodological basis of world water balance investigation and computation. IASH Publication, no. 148
UN-Water (2010) UN-water task force on indicators, monitoring and reporting: final report. In: Monitoring progress in the water sector: a selected set of indicators. Word water assessment programme
Van der Gun J (2012) Groundwater and global change: trends, opportunities and challenges. WWDR4, UNESCO
Vitousek PM, Ehrlich PR, Ehrlich AH, Matson PA (1986) Human appropriation of the products of photosynthesis. BioSci 36(6):368–373 (Jun, 1986). University of California Press, American Institute of Biological Sciences
Wada Y, Beek LP, Sperna-Weiland FC, Chao BF, Wu YH, Bierkens MFP (2012) Past and future contribution of global groundwater depletion to sea-level rise. Geophys Res Lett 39:L09402. https://doi.org/10.1029/2012GL051230
Wada Y, Lo MH, Yeh PJF, Reager JT, Famiglietti JS, Wu RJ, Tseng YH (2016) Fate of water pumped from underground and contributions to sea-level rise. Nat Clim Change 6:777–780. https://doi.org/10.1038/nclimate3001
Worldclim.org (2011) http://www.worldclim.org/ (consulted on 2 June 2011)

Chapter 2
On the Water Security Concept: State of the Art

Contents

2.1 Introduction

Water security involves technical, economic, environmental, social and legal aspects. The status of public property (or the public domain) of water resources, written into most legislations, allows the public authorities to achieve public utility objectives. This status has been the basis of water authority centralization, and has promoted mobilization policies that put water at the service of economic and social developments. This is particularly the case in water scarce countries where, in some regions, water management programs have led to a near total control of surface and groundwater resources, sometimes with excessive levels of water withdrawal, which threaten the resources sustainability.

The economic and demographic transformations that have taken place in contemporary times have had important consequences on water resource management: water needs and uses have diversified and increased considerably and the effects of pollution release have become increasingly threatening the integrity of water environments. At the same time, societies have developed and their concerns with regard

M. Besbes et al., *National Water Security*,
https://doi.org/10.1007/978-3-319-75499-4_2

to environmental issues have become concrete. In order to assimilate these new challenges, the legal systems of water have adapted in various ways to the new conditions created by changes in habits, habitat, industry, agriculture, leisure.

This chapter develops the essential elements of a framework that conceptualizes water security at the national level. It starts by a historical overview that traces the progress of the legal status of water and of the concepts and methods employed in addressing water security issues and attempts to place the reader in tune with current trends, debates and sometimes contradictions that dominate the issue of water on a global scale. Then the National Water Security concept is developed to integrate in particular the constraints and challenges of arid countries.

2.2 Genesis and Growth of the Water Security Concept

Water resources, essential for life and economic development, are inextricably linked to the prosperity and welfare of individuals and populations. Water may in return constitute a direct factor of insecurity or peace deterioration through its potential of tensions outbreak, conflicts and disputes, especially as a substantial portion of surface water and groundwater is shared by two or more communities or nations. Consequently, the balanced relationships between close communities or neighboring nations, are closely linked to efficiency of local regulations, national laws, international agreements, which allow equitable access to shared water resources.

Today, the international community in its various manifestations has always regarded access to water and its equitable sharing as factors of stability and peace throughout the world. In addition, water is intimately linked to environmental issues, which are increasingly recognized as a major concern for the international community. At the national level, the implementation of water safety is the responsibility of the authorities. They are called upon to develop appropriate policies, laws and regulations, and to allocate the necessary budgets for the implementation of strategies and programs to improve water security.

Measurement of the level of water safety has fueled an abundant literature. The debate was initiated by applying an agronomic concept to human needs: for a given country, the indicator of "water stress" (Falkenmark et al. 1989) represents the threshold of renewable water resources per inhabitant, available for domestic, industrial, agricultural and environmental needs. Other researchers have developed water stress indicators that are better able to characterize actual pressure on water resources, taking into account the current water withdrawals. However, while these two conceptions of water stress (per capita water availability and resource harvesting rates) reflect anthropogenic pressure on the resource, they do not reflect the importance of the quantities that are effectively consumed, and the share released into the natural environment likely to be recycled into the water resource. Similarly, the water stress indicators do not concern the water resource involved in rainfed agriculture and that used by natural vegetation (green water), which can represent much more important quantities than surface and underground water resources employed across the country (blue water).

Water security is at the same time an internal and international issue. This point is developed from the two first points of view: (i) The evolution in the juridical status of water: what's the legal qualification of water at national and international levels and how water issue is-it related to the general principles of international law? (ii) Water security as an international issue: what is the international community action in addressing the water issue? (iii) The third point is devoted to the right to water: the meaning and scope of the right to water are specified and an attempt is made to analyze the progress made in its implementation. The following sections are devoted to water security and water scarcity assessments. By presenting key issues of traditional water scarcity assessments, we try to consider the most prominent progress in the direction towards more holistic approaches which goes beyond conventional assessments of water availability and water scarcity, to include the multiple dimensions of water security: (iv) Water Scarcity and Water Stress indicators, (v) Water dependency and the holistic water security, (vi) The Water Security concept.

2.2.1 The Water Status for Individuals and Nations

Water qualification regarding national laws

Continental law (Civil Law) of the Romano-Germanic tradition and Anglo-Saxon law (Common Law) recognize the common character of water resources. In Roman law, water is originally referred to as "common thing" (res communis), a concept which designates things which, because of their essential value for life, are removed from the regime of property. Similarly, the Common Law has traditionally conceived that there can be no ownership rights over surface water and groundwater. According to the Islamic tradition, water is a gift from God: one is bound to share it and no one has the right to appropriate it exclusively. But if the concept of water commonality is anchored in societies' heritages, most legal systems establish uses rights. The Civil Law as well as the Common Law grants rights of use to those who have access to water through their funds. The right to water is then linked to land ownership, called "riparian rights", with rules of use (Pozzo 2000). Thus, in the northeastern United States, where water is abundant, "riparian rights" prevail whereas in the southwest, where water is scarce, privilege is granted to acquired rights or "prior appropriation", (Cárdenas 2007). Equivalent formulations can be found in the "Medjellé" (Ottoman Civil Code), which distinguishes "open access" (mubah) waters from waters where certain communities may have rights (musha). In this case, the chronological priority is applied and Medjellé recognizes "prior appropriation", (FAO 1977).

Alongside the principle of access to the common thing, Roman law introduced the notion of a public thing (res publica), which designates "things out of the commerce, made available to all" and whose control is under the public power. This status applies to navigable rivers. The civil law has consecrated the authority of the State and strengthened the appropriation of water by the public authorities by establishing the domaniality of the important rivers; the latter are then classified as "public goods". The public good differs from the public thing in that the good has a pecuniary value. As a good, water therefore has an economic value which, when the resource is scarce or limited, creates competition between different uses. The public authority intervention is then necessary to reconcile uses, define rules and access conditions, and achieve ends of "public utility". This status was the basis of water authority centralization, and promoted mobilization policies that put water at the service of economic and social development.

By qualifying water as an "element of the common patrimony of the nation", the French law of 3 January 1992 was the first to recognize water as a legal entity free from any reference to land. The concept of patrimony, which the French legislature uses in other jurisdictions, gives water a qualification that goes beyond the notion of ownership (public or private) and incorporates a moral element of responsibility with regard to past and future generations (Gaonac'h 1999).

Water in the international law

While significant progress has been made at the national and regional levels, the legal status of water at the international level has not made the same progress. At the international level, water does not have a special status and the general status of natural resources is applied by default to water resources, (Paquerot 2004). Thus, international law subjects water, like all natural resources, to the principle of free trade. This legal concept does not integrate the social, environmental and cultural value of water and does not allow for specific regulation of water resources at the international level. The status of water is still the subject of a broad debate aimed at promoting the construction of an international legal regime compatible with the multiple challenges of water and the environment.

To this regard, the concept of sustainable development has quickly become established as one of the international community objectives. Its implications for the exploitation and management of natural resources are considerable. Sustainable development has been defined in many ways, but the most frequently quoted definition is from "Our Common Future", also known as "the Brundtland Report": "Sustainable development is development that meets the needs of the present without compromising the ability of future generations to meet their own needs". Inserted among the Millennium Development Goals set by the United Nations (UN) Member States and conceived as a break with other modes of development which has led to considerable social and ecological damage, it aims at the implementation of viable schemes capable to reconcile the three stakes of the human activities: economic, social, and environmental.

At the global level, it is recognized that good water management is a decisive element in achieving sustainable development objectives, with the need for political commitment on the part of public authorities, broad public support and the important contribution of the civil society. At the regional level, water management is one of the major concerns of the arid and semi-arid countries, at the center of all sustainable development, while in number of these countries, water withdrawals approach the order of magnitude of resources. As water scarcity exacerbates the water-related tensions between sectors and between water users or States, specific programs, supported by international cooperation and solidarity are launched to provide appropriate responses to the critical water issues in arid countries.

2.2.2 *Water Security as an International Issue*

The international community action in the area of environment and water management is important at both the cognitive and institutional levels. In this regard, the UNESCO's (United Nations Educational, Scientific and Cultural Organization) "Arid Zone Hydrology Program" can be regarded as the precursor of the scientific globalization of water issues. On the other hand international conferences, forums, summits, conventions contributed to the formulation of institutional frameworks, strengthening national capacities and implementation of water management programs.

The "Arid Zone Hydrology" program of UNESCO

The work of Batisse (2005) brings a remarkable historical light on UNESCO's contribution to the progress of water knowledge and water vision. In 1948, the General Conference in Beirut recommended the establishment of an international institute for the arid zone, which evolved into a "Research Advisory Committee on Arid Zone". In 1951, is born the Arid Zone Program of UNESCO, in which the issue of water resources occupied a prominent place. The first visible result of the program was held in 1952 with the Ankara Conference, the first global scientific meeting on the Hydrology of Arid Zone. The Hydrology was then a major scientific project of UNESCO, by which the international scientific community would contribute to the solution of concrete population problems.

Thus emerged the will to "encourage the establishment of more numerous databases on water resources, and foster international exchange of information and experience acquired in this field". And in 1960 UNESCO put the Water Security issue on the agenda of its fields of interest, "seriously worrying the enormous difficulty in meeting water needs, both domestic and industrial, a rapidly growing world population; emphasizing the global nature of the water resources issue and the lack of knowledge on the subject".

This concern was materialized by a resolution aimed at developing international collaborative research and training programs in the field of scientific hydrology, appointing a commission of experts, whose rapporteur was Lassâad Ben Osman, the Tunisian rapporteur of the Arid zone Advisory Committee, to develop an Action Program. This initiative found its culmination with the launch of the "International Hydrological Decade": 1964–1974, followed by the International Hydrological Program (IHP) and its different phases.

The Decade constituted a formidable promotional movement for Hydrology, in the triple field: operational, scientific and cognitive. At its start, it was realized very quickly that there was a lack of water experts and hydrologists: unprecedented efforts were then made by UNESCO authorities to initiate training at all levels. The start of the Decade corresponded with the arrival on the scene, with the national independences, of all young people and new generations of hydrologists, promoting their professional and scientific development by mixing of experiences which accompanied this first scientific globalization enterprise.

One of the relevant issues, concerning both the advancement of knowledge on arid zone hydrology and decisive changes in the mode of groundwater exploitation in water scarce regions, was the launching, under the aegis of UNESCO, of the Northern Sahara Water Resources Study project (NSWRS), which constituted the first attempt of knowledge and common management, by Algeria and Tunisia, of a very large transboundary aquifer system. The completion of the project, in 1972, impulsed methodological and practical decisive breakthrough in the understanding of water security in the Sahara arid regions.

As for the International Hydrological Program (IHP), it still constitutes the only intergovernmental program devoted to research in the field of water, management of water resources, education and capacity building. This program is being implemented in phases of six years, allowing it to evolve according to the needs of a changing world.

Water and the international conferences

The International Conference on "Water for Peace", held in Washington in 1967, marked the beginning of international cooperation in the water sector, noting that water issues can only be addressed by an approach that goes beyond individual nations and that it is therefore necessary to develop exchange of knowledge, experiences and information and to promote technology transfer, training and development of human resources in the field of water. But water is also concerned with issues relating to the environment. The first world meeting that has placed ecology on the international agenda was the UN Conference on the Human Environment held in Stockholm in 1972, which recognized the right to the environment as a "fundamental right".

In the wake of Stockholm was held, in 1977 in Mar del Plata, the UN Conference on Water, dedicated to issues on assessment and use of water resources, and whose most striking result was the launch of the Decade for drinking water and sanitation. Proclaimed by the United Nations in 1981, the Decade has set an ambitious goal of ensuring all people access to safe drinking water and basic sanitation by the year 1990. Despite significant positive results, the decade results remained below expectations: in 1990, 1.3 billion people lacked access to safe drinking water and 2.6 billion people did not enjoy adequate sanitation services. The end of the decade was marked by the organization of a meeting held in Montreal where hundreds of organizations concerned by a better distribution of water have adopted what would become the Charter of Montreal, which proclaims the right of access to drinking water as a basic human right. The Decade has raised awareness of the complexity of water-related projects and technical and financial difficulties associated with them: the financial reasons for the relative failure of the Decade reinforced the perception of the economic value of water dedicated at the Dublin Conference in 1992, which sets out the "Dublin principles".

The United Nations Conference on Environment and Development (UNCED) in Rio in 1992 (Earth Summit) has contributed to deepen and strengthen the principles of governance and water resource management. The action plan adopted (Agenda 21) defined the guiding principles of management, where water is presented as "a natural resource and a social and economic good."

At the New York Millennium Summit in 2000, the international community pledged to achieve the eight Millennium Development Goals (MDGs) to reduce poverty by half by 2015. The 7th Goal entitled "Ensuring environmental sustainability" recommended among others things, to reduce by half the population without access to safe drinking water and basic sanitation.

Two years later stood the World Summit on Sustainable Development in Johannesburg in 2002, which called for the development of rational plans and integrated water resources management at various scales, while reaffirming the Millennium Development Goals. Halfway through the MDG deadline, the 2008 WHO report paints a mixed picture of the situation of water supply and sanitation equipment: MDG on access to clean water are about to be realized but with huge regional and local disparities. Of the 884 million people in 2008 who did not have access to improved sources of supply, 40% are in sub-Saharan Africa. The situation is more delicate for sanitation where 2.6 billion people still do not have improved sanitation and trends indicate that this figure is set to increase.

With the resolution of the UN General Assembly in July 2010, the right to water is formally recognized by the world's highest political authority. Non-binding, this resolution has no legal value, but is considered as having a moral value and an important historical fact. Now, the right of access to water principle is introduced into the World Heritage principles. It should eventually lead to strengthen commitments to greater water solidarity at the different scales: national, regional and international.

The World Water Forum

It is under the auspices of the World Water Council that every 3 years since 1997, is held the World Water Forum, a space of reflection and debate on issues related to water. The success of the World Water Forum is indisputable and confirmed by the interest shown by the international community at this event where local authorities, civil society, experts and decision makers combine to make proposals and take action on the political water agenda. The declaration of the first Forum in Marrakech, 1997, announces management principles: "recognize the basic human need to have access to safe water and sanitation, establish an effective mechanism for the management of shared waters, support and conserve ecosystems, encourage the efficient use of water…". In 2000, the Declaration of the "World Water Vision" presented at The Hague Forum has not brought new perspectives. It describes the water as an element "essential to the life and health of humans and ecosystems and a fundamental condition for the development of countries" and calls for "good water management: ensuring good governance, particularly by participation of the public and key stakeholders."

The following forums were held in Tokyo (2003), Mexico City (2006), Istanbul (2009), Marseilles (2012) and Daegu/Gyeongbuk (2015). The final declarations reiterate the general principles: Integrated Water Resources Management (IWRM), local participation, sharing information, management of water related risks. The Istanbul Declaration is unusually detailed on economic issues by introducing the "3T" concept (Tariffs, Taxes, and Transfers) and recommending sustainable cost recovery. It also stresses the need for cooperation between private and public operators. The Daegu/Gyeongbuk forum introduced the EWI (Economically Water Insecure) classification. While physical water scarcity refers to inadequate access due to high water stress, EWI describes a situation where access is limited despite low water stress levels. It is defined as "the condition of lacking water security due to economic factors such as lack of investment in water infrastructure, water data monitoring and low water management."

2.2.3 Water Security and the Right to Water

Taken as a whole, the different legal systems give water juridical statuses that differentiate it from other natural resources and that take into account, in different ways, the obvious basic rights of access to water and of equity in its distribution. The notions of the common good, public good or common patrimony of the nation have been used to describe a resource essential to life. On the other hand, the international community has been very active in deepening and entrenching human rights including the rights to water and to sanitation. General Comment No. 15 on the right to water was adopted by the UN Committee on Economic, Social and Cultural Rights in November 2002 affirms that "the human right to water entitles everyone to sufficient, safe, acceptable, physically accessible and affordable water for personal and domestic uses".

The right to water is basically about the quantities needed to meet basic human needs. It also relates to sanitation. This therefore concerns relatively small quantities in comparison with the quantities of water used for economic activities. A distinction should be made between the right to water, which refers to the access to safe drinking water and sanitation, from water rights which refer to the use of water for specific purposes (irrigation, electricity production, industrial use…). In general, water rights are governed by texts defining the beneficiaries of these rights, the quantities allocated and the conditions of use.

Some countries, such as South Africa, Congo, Ethiopia, Gambia, Uganda, Zambia and Ecuador, have for more than a decade, explicitly introduced the right to water in their constitution. More recently, the Tunisian Constitution of 2014 explicitly enshrines the right to water and equally explicitly imposes the duty of the State and society to preserve the resource and guarantee the rational use of water. In a general way, contemporary water management practices evolve towards holistic approaches that consider the entire hydrological cycle and the different uses that influence water regimes. The European Water Framework Directive (2000/60/EC) obliges the member States of the European Union to implement these principles.

Furthermore the European Union opted in 2014 to exclude the water supply and water resources management from the rules governing the European internal market. Recently, Slovenia included in 2016 the right to drinking water in its constitution, insisting that this resource cannot be privatized. By doing this, Slovenia became the first country in Europe to give to the right to water a constitutional status.

In general, the adoption of the right to water obliges States to devote the required funds for its implementation. This translates into a form of solidarity between citizens and regions to make to water accessible to the entire population. In Tunisia, for example, uniform pricing of drinking water distributed in urban areas is applied thought the entire country, irrespective of the source of water (surface water, groundwater, desalination) and whatever are the development methods adopted to provide water (mobilization, transfers) and the various water treatments applied to make it safe to drink.

2.2.4 Water Scarcity and Water Stress Indicators

The term of "security" recovers many different meanings ranging from the concept of "human security" aiming at protecting individual people and their community to many other security issues including food security, energy security and access to natural resources, in particular to water resources. The accessibility of population to safe water and the provision of sufficient water of adequate quality to meet all sectorial water demands involves physical water availability as well as social, economic and political factors. Water scarcity is becoming a current focus of considerable awareness in relation to population growth and environmental issues. Much of the conventional measures of water scarcity are based on average water resources availability emphasizing the physical water scarcity. Applied at a national scale, traditional water scarcity assessments have little sense in characterizing water scarcity at local scales since water scarcity is determined not only by supply availability but also by the demand development. This makes it difficult to determine the nature of water scarcity; is it due to real scarcity in the physical sense or is it related to insufficient water management?

The Water Scarcity

Water scarcity occurs where there are insufficient water resources to satisfy economic and social uses, as well as ecosystems requirements. The concept of water scarcity is not only limited to quantitative aspects; it embraces also the quality of water. Thus water scarcity, firstly due to low water availability, is accentuated by the increasing of water demand, by water resources depletion and by water quality deterioration.

Water scarcity can be physical or economic. Physical scarcity occurs when there are not enough resources to meet all demands. Economic water scarcity is caused by a lack of investment capacities in water facilities to satisfy the water demand. Arid regions are most often concerned with physical water scarcity, but water scarcity problems also appear in number of water abundant regions lacking hydraulic infrastructures to mobilize resources.

The notion of Water stress derives from the application of an agronomic concept to human needs. Water scarcity involves water stress and water shortage or deficits which may represent important obstacles to the socio-economic development of water scarce areas. Both physical and economic scarcities highlight aspects of anthropogenic pressure on water resources.

The Water scarcity indicators

Several initiatives and methods have been developed to assess water scarcity. Most are based on annual average data and applied on a national scale for which data is readily available. Two indicators are widely used for defining and measuring water stress: the first evaluates the per capita water availability; the second determines the withdrawal to water resources. Each of these indicators incorporates a specific scaling with conventional households which characterize water stress and water shortage situations.

One of the most commonly used measures of water scarcity is the "Water Stress Indicator" (WSI) of Falkenmark (Falkenmark et al. 1989). The WSI measures the water resource abundance or scarcity by comparing the available freshwater resource to the population, considered as the essential factor of the water demand. Thus the Falkenmark indicator, which appears as a "water crowding index" lends itself well to prospective studies based on demographic projections. The "Water Stress Indicator" of Falkenmark differentiates four stress levels based on per capita water availability as: (i) No stress (>1700 m^3/year/capita), (ii) Stress (1000–1700), (iii) Scarcity (500–1000), and absolute scarcity (< 500 m^3/year/capita), where estimates of water resources are typically derived from mean annual runoff.

Another well-known water scarcity indicator which is equally easy to determine and interpret was proposed by Raskin et al. (1997). This indicator characterizes the water stress by evaluating the ratio of annual withdrawals to available water resources. The Withdrawal to Water Resources (WWR) or Use-To-Resources Ratio can thus be viewed as a water resource vulnerability index used to better capture the concept of use instead of demand, defining scarcity as a percentage of availability and classifying countries on the basis of the level of water extraction. According to this approach, a country is water scarce if annual withdrawals are between 20 and 40% of annual water resources, and severely water scarce if withdrawals exceed 40%.

Both WSI and WWR are today the most widely used indices for measuring water scarcity because of the relative ease with which these indicators can be calculated using annual averaged data at country levels. Both of the two indicators are intended to characterize the scarcity level using different methods that highlight important aspects of water scarcity, and thus complement each other, but none of them reflect the conditions for effective access of populations to water. For a given level of resources and population, a country is certainly less vulnerable if it has already mobilized, without reaching the limits of exploitation, a sufficiently large share of its resources, which ensure a level of water sufficiency compatible with the standard of living of the populations. In this regard, Besbes (1998) defined a Water Comfort Index (WCI) which measures the amount of water effectively used by the population. Indeed, it is difficult to characterize water scarcity without specifying the amount of water effectively made available for population and the extent to which it can meet its real needs.

Another central aspect of water scarcity is related to the level of water resources abstraction from the natural water cycle which has to ensure environment sustainability. Sustainable water abstraction has to be carried out within the limits of the natural resources and its ability to ensure specific environmental water demand in particular in safeguarding ecosystems and biodiversity.

Further developments on Water Scarcity Indicators

The main inadequacies in estimating water scarcity indicators refer to the inability of the indicators to represent the different components of water resources and water needs balances (or unbalances) over different geographical and time scales.

The International Water Management Institute (IWMI) developed a measure of water scarcity at national scale in order to introduce future adaptive capacity of the country to deal with scarcity accounting for its existing water infrastructure and for its potential to increase water supply and improve water management (Seckler et al. 1999). According to this approach, water scarce countries have been classified in two groups depending on the degree of renewable water resource available for human needs: (i) physically water scarce countries; (ii) economic water scarce countries. The latter have sufficient water resources to meet projected requirements but need to embark on massive water development programs to actually utilize these resources and make it available to people in the future.

A more comprehensive indicator measuring water availability and quality, at the household and community levels, is proposed by Sullivan (2002), in the form of a Water Poverty Index (WPI). The WPI addresses physical water availability/scarcity and human and ecosystem demands/uses. It combines various water-related issues (access to water; water quantity, quality and variability; water uses for domestic, food and productive purposes; capacity for water management; environmental aspects and ecological integrity) to capture the complexity of the water situation of a community or country. Sullivan et al. (2003) strongly argue that the WPI is a systematic approach that "provides a means of understanding the complexities of water issues by integrating the physical, social, economic and environmental aspects, and by linking water issues to poverty". These authors see great potential in the use of the WPI, presented as a powerful tool for decision-makers and for local communities to determine priorities and to justify choices. Nevertheless, and despite the advantage of its comprehensiveness, the WPI is hampered by its complexity and lack of intuitive understanding in comparison to easy-to-understand indicators such as the Falkenmark index. This would seem justified today, judging by the actual use of water scarcity indicators. Despite the significant shortcomings of the water scarcity indexes based on water stress thresholds, and notwithstanding the progress achieved in the development and updating of the methods and approaches of assessing water scarcity at different scales, the water scarcity indexes based on water stress thresholds and in particular the Falkenmark index, are still however largely used to characterize water scarcity in many studies on climate change or on water security. On the other hand the simple water scarcity indicators are undoubtedly insufficient to duly cover all aspects of water and the issues related to it.

2.2.5 *Food Dependency and the Holistic Water Balance*

Water scarcity related to the water-food nexus

The basic water needs represent only a small part, few percent, of the overall water demand. In contrast, the part of water demand involved in foodstuffs production is relatively high. Many works have been devoted to the evaluation of the quantity of water required in foodstuffs production. In spite of their disparities, data resulting from research provide edifying information on the relation between water and foodstuffs production. The water needs of agriculture and food production are high and may vary due to changes in dietary habits associated particularly with an improved standard of living and purchasing power.

Disregarding the origin of water used in agriculture, the production of 1 kg of cereals in temperate areas needs approximately 1 m^3 of water. However, depending on how much feed is given to the animals versus animals that graze on rainfed pastures, one kilogram of meat requires much more water: 15–20 m^3.

According to Zimmer and Renault (2003), the water consumed by crop production represents, on average at global scale about 2 m^3/day/capita. Taking into account the contribution of grasslands, the total water involved in food production is expected to reach 3 m^3/day/capita. Moreover, the various data clearly show the inequality between the water amounts involved in population feeding in the different regions of the globe. Countries with a diet relatively high in animal product intake (North America and Europe) use at least 3 m^3/day/capita of water to nourish their populations; while countries with a lower diet in animal product intake (Asia and Africa) use much less with values of about 1.2 m^3/day/capita.

The important quantities of water involved in food production indicate that a full understanding of water issues should consider the structure of agricultural production. Adequacy or scarcity of water resources depends mainly on the role that society assigns to irrigation and the place it is supposed to play in development policies in general and in agricultural policies in particular. Moreover, this immediately means that, in many arid and semi-arid countries, the implementation of agriculture policies based on modern irrigated agriculture involves a physical limit which is that of water resources.

Water and food security

An important factor underlying the general notion of 'security' at national scale refers to the ability of countries to meet their own needs. This is generally referred to as 'self-sufficiency'. 'Food self-sufficiency' and 'food security' designate completely different concepts. According to Rome Declaration on Food Security and World Food Summit Plan of Action: "Food security exists when all people, at all times, have physical and economic access to sufficient, safe and nutritious food to meet their dietary needs and food preferences for an active and healthy life" (FAO 1996). In contrast, food self-sufficiency is defined as being able to meet consumption needs from domestic production rather than by imports. The food demand increase in many water scarce countries has triggered the need to organize and regulate the food trade and led to gradually moving from self-sufficiency objectives to food-security concern. Given the large amount of water resources involved in food production, it might be questioned to what extent agri-food policies may influence the development and use of water resources.

The holistic water resources effectively used by the population can be defined by the following four components; (i) available water resources mobilized from surface and groundwater systems; (ii) soil water resources involved in food production (crops production, pastures); (iii) water resources produced by non conventional methods (Treated Waste Water Reuse, Desalination); and (iv) Virtual Water flows related to trans-boundary trade. The first two components can be considered as part of the natural water cycle. Along with the third component, this represents what is produced and used locally; the fourth component of holistic water resources represents the cross border flux of virtual water the balance of which represents the difference between all the water needed by the population and all the water used by it.

If positive, the virtual water budget means that, in balance-sheet terms, the holistic water resources used locally are not sufficient to meet all water demands and the deficit is met by relying on imports. In the context when the water availability becomes a limiting factor because of the resource potential, the direct water demands for urban, industrial, touristic uses have in general an obvious priority in the resource allocation. The increased allocations for direct water uses can only be made at the expense of agricultural use. It results in sheet-balance terms so that the virtual water contribution shall correspond to the water volume needed to meet food demand.

The Water Dependency Index

The comprehensive water balance expresses the amount of virtual water associated with food products trade and defines a "Water Dependency Index" (WDI), which represents the part of net virtual water in the total food demand water equivalent. This assumes that the allowance in Blue Water to irrigation must adjust to the available water once the direct needs insured.

The Food Demand Water Equivalent (FDWE) includes water equivalent of Agricultural Production consumed on the local market and the water-equivalent of agri-food imports. The water equivalent of agricultural production includes green water-equivalent and blue water equivalent.

The "Water Dependency Index" (WDI), defined by Besbes et al. (2002) and (ITES 2011), represents the net equivalent of Virtual Water volumes (Imports-Exports) within the total food demand and is expressed as: WDI = (IMP–EXP)/FDWE.

If one refers to international literature, the concept of water dependency, as defined by FAO (2003), relates only to blue water; it expresses the external renewable water resources (originating outside the country) as a percentage of the total renewable water resources (internal and external). This definition has been largely used by the scientific community as well as by international organisations.

Based on water footprint concept, Hoekstra and Mekonnen (2012) defined the 'virtual water import dependency' of a nation as 'the ratio of the external to the total water footprint of national consumption' where total 'water consumption' refers to the 'water needed for the production of the domestic demand for goods and services'. The indicator is conceived to reflect the extent to which a country relies on imports of water in virtual form. The results reported by Hoekstra and Mekonnen (2012) on water dependency give, as it may be expected, high values for water-scarce countries (like Jordan 86%, Israel 82%, Yemen 76%, Lebanon 73%). These results reveal however some striking points; in particular, some water-rich countries such as Italy, Germany, the United Kingdom, and The Netherlands have surprisingly high water dependency indexes between 60–95%.

By relating the Water Dependency Index to agricultural water (Besbes et al. 2002; ITES 2011), the WDI attempts to go beyond the appraisal of the water dependency level of nations to specify the balance sheet items related to the national food demand. As the net equivalent of virtual water represents the difference between the total food demand water equivalent and the total food production water equivalent, the WDI could be more explicitly detailed in order to bring out the different contributions to food production: "Blue Water" referring to the use of ground and surface water as well as non conventional water resources, "Green Water" referring to the water reserves of the soil effectively used in crop production or into direct grazing, and "Virtual Water" referring to the flux of the "net virtual water import". The objective is to consider the extent to which greater value for all water resources could be achieved.

As the major part of water resources is directly or indirectly used in food production, the WDI related to food balance is in itself sufficient to reflect the National water security by measuring the level to which a nation relies on foreign water to ensure its food demand. This indicator could be consolidated by financial indicators, for instance the coverage rate of the agri-food trade balance. The improvement of the food security of a country expressed in terms of WDI will depend on the capacity of the country to improve food productivity either in the irrigated sector (Blue Water, including non conventional water resource) as well as in the rain-fed agriculture and direct grazing (Green Water). From this point of view, the WDI appears as a major decision-making tool for sustainable water resources management. It is also a learning tool as well as a 'discussion-support' tool that provides a common platform for coherence of the activities of different actors and stakeholders.

2.2.6 The Water Security Concept

Water Security and Integrated Water Resource Management concepts

The multidimensional character of water has led to sectoral approaches in the organization of water administration. Integrated approaches to water management provide a framework to reconcile efficiency, equity and environmental sustainability. It has become a must for all water policy makers to include the holistic dimension of water resources as well as social and environmental impacts. Nevertheless, due to concerns for national sovereignty, national-level integrated water resource management (IWRM) policies can sometimes be developed without sufficient consideration of the transboundary nature of international watercourses and aquifers. They may also not adequately consider existing local and traditional practices tasked with water resources management and allocation. The cross-border nature of water basins underlines the need for international cooperation and transboundary institutional structures. On the other hand, national water management should enhance its relevance at the local level, taking into account socio-economic, political and cultural factors and promoting community participation.

Recent years have seen the emergence of new and more open approaches that address water issues in integrated perspectives, combining technical, economical, environmental and social dimensions. These increasingly holistic approaches refer to the integration of quantitative and qualitative aspects of water resources and to the conciliation of the socio-economic function of water resources and its role in the protection of natural environment. Recent achievements has emerged an all-inclusive and holistic concept: the water security. The term was first used in the Ministerial Declaration of the 2nd World Water Forum (WWF) in The Hague in 2000. Since then the term has been widened and deepened.

Latest reference materials highlight the diversity of approaches to water security. The books by Lankford et al. (2013) and Pahl-Wostl et al. (2016) are both conceived in the form of authors' contributions to reflect the diversity of approaches to water security. Built on a pluralistic discourse, both books allow for debates on the definition, usage and interpretation of water security concept. Topics covered include water availability, human needs (minimal access to water, agricultural productivity, food security), human vulnerability to hazards (floods, droughts, hydrologic variability, contamination and terrorism, infrastructural security), and sustainability (climate change, ecosystem needs) as well as interdisciplinary and nexus linkages.

Water security covers different meanings and definitions, one of the most cited being that of Grey and Sadoff (2007): 'availability of an acceptable quantity and quality of water for health, livelihoods, ecosystems and production, coupled with an acceptable level of water-related risks to people, environments and economies.'

Water security embraces both the productive potential of water and its destructive impact. It also includes economic and social aspects as well as ecological and political issues. All these dimensions of water security cannot be tackled on a sectoral or local basis, as they are structurally linked to other security issues at multiple scales: global security, national security, human security, food security, and energy security, incorporated into the new concept of water, energy and food security nexus (WFE-Nexus). Therefore, the term "water security" attempts to capture the various dimensions of water-related issues and involves many security areas.

Reviewing the emerging literature on water security, Gerlak and Mukhtarov (2015) discussed the relationship between water security concept and the more traditional dominant concept of IWRM. They found that the IWRM is "integrative in terms of science rationality and analytical planning" while, water security highlights "issues of ethics, values, the human dimensions of water management and the sense of urgency in dealing with water". They argued that the two concepts are complementary rather than conflicting so that the coexistence of the two concepts provides solution for global water community and for the future of water governance. Their research concludes that it would be beneficial in implementing national water planning programmes, to see who, among the major actors (donor-led projects and decision making in international organizations) are making use of IWRM and water security and what strategies are being used in promoting these concepts.

Water security in national and international organizations agendas

Both national and international organizations have taken an important position in order to promote water security and in setting up international cooperation programs, taking advantage of the positive momentum generated by international conferences. There is consensus that water security is built on a comprehensive and integrative approach that captures a range of longer-established concepts (national security, political security, human security, food security, energy security) and incorporates both the human side and the ecological side of water issues, (Zeitoun 2011). Water security can only be achieved if it is conceived with a sustainable development vision that integrates climate change consequences.

Therefore, the involvement of national and international organizations (Agencies within and outside the United Nations system, international and regional development banks, professional associations, non-governmental organizations, private sector) in water security is promising in moving from conceptual formulation to operational implementation because their interventions are directed toward action-oriented projects and programs in their respective fields of activity. Depending on their areas of interest and expertise, the international organizations have contributed to deepening the debate by putting forward certain practical and operational aspects. In this field, comprehensive information is given and relevant analysis and strategies regarding the different water security issues are developed with, in some cases, instructive experience feedbacks (Global Water Partnership 2010; AFD 2015).

It is common, in security theory to distinguish between hard and soft security issues. According to Fatić (2002), hard security is depth based, it corresponds to security threats that are primarily external, and involves a substantial ability to respond, including the use of military force; while soft security is width-based and corresponds to primarily internal or trans-border threats that involve policy-priority response capacity. Many national and international organizations involved in water security and in the first place, the UN Agencies, advocate increased cooperation to identify and resolve tensions before they generate into conflict. More specifically, there appeared to be general consensus that although real potential exists for conflict over water, water tensions can also offer potential for cooperation (Bigas 2012). Moreover water issues may facilitate the conversion of conflict to successful trans-boundary and regional cooperation and may contribute to preventing and resolving water conflicts, (Bigas 2013; Houdret et al. 2010).

The international cooperation is also essential to provide financial and technical support to achieve water security objectives and priorities. According to the Task Force on Water Security and Sustainable Growth (Sadoff et al. 2015), the total cost of water insecurity for the global economy, amounts to $500 billion. Taking into account environmental impacts, the cost of water insecurity would represent 1% of world GDP. The financing of water security is thus a key issue which calls for the mobilization of international funds and resources that may go beyond national financial possibilities, in particular those of developing countries.

Number of financial providers and international cooperation agencies are getting more involved in investing in water security. Some of them developed investment policies, standards and procedures to govern investment in water programs. The Asian Development Bank (Asian Development Bank 2016) considers that "water security is more than just providing sufficient water for people and economic activities. It is also about having healthy aquatic ecosystems and protecting us against water-related disasters". In order to assess overall national water security of Asia and the Pacific countries The ADB developed a water security framework based on five key dimensions (KDs) for household, economic, urban environmental, and resilience to water-related disasters.

According to the World Bank (2005) "Water security is achieved when water underpins economic growth, rather than undermining it". Under this vision, investments in water have to "focus on growth enhancement, rather than unfulfilled basic needs and risk mitigation". However the World Bank highlights that this water security funding framework can hardly be considered when basic water security is not yet achieved. This means that the achievement of water security requires an acquisition of a Minimum Platform Investment (MIP) under which water may compromise growth overall. The World Water Council (2015) argues that investing in water security is an essential condition for economic growth and for breaking cycles of poverty. This will be particularly significant in the poorest countries where economic structure and resilience "affect the minimum level of institutions and infrastructure necessary for water security". To this regard, the Mexico declaration (4th WWF Mexico 2016) of African Ministers Council on Water declared "African countries need to invest in water infrastructure up to the level where they can, in order to achieve a self-sustaining auto-induced growth to eradicate poverty and achieve sustainable development". This is an area in which the African Development Bank (AfDB) and the African Water Facility (AWF) are heavily engaged; in particular in water programs leading to greater development impact, notably those related to the achievement of the water and sanitation MDGs targets. Another important aim of MDGs in addressing poverty is related to food and nutrition problems. This also requires investment in water infrastructure to develop agriculture and to enhance the welfare of households and communities, (FAO-WWC 2015).

Naturally, one of the most notable aspects of water security is related to the access to water for the basic human needs (drinking, cooking, bathing, sanitation and hygiene). To this regard, the recognition of right of access to drinking water and sanitation by the UN General Assembly and the UN Human Rights Council has constituted a progress of enormous importance to humanity.

2.3 The National Water Security Concept

2.3.1 *"Hard Water Security" Issues at National Scale*

Hard water security issues are about managing, preventing and mitigating risks. Water security is achieved when all risks and dangers related to water management are avoided. As water is an important strategic resource, all "hard security" issues related to water management fall within the area of national security concerns. At "National" scale, water security becomes a central political issue for which public authorities have to develop appropriate strategies to secure supply for all uses and promote human well-being through socio-economic development. The water hard security involves both internal and external security components.

Internal hard security issues related to water

Internally, many countries developed tools to enhance the security of drinking water systems. Some countries attach special importance to the protection of water resources and installations against possible terrorist acts (US EPA 2011), while others are introducing equipment to prevent emergencies in the drinking-water supply of large metropolises in the event of a serious crisis (DRIEE Ile de France 2014).

Generally speaking, drinking water supply systems can be threatened by pollution accidents, terrorist attacks, and natural disasters. Water authorities at central and local scales should establish a framework to prevent different types of accidents, to develop emergency response capacity and to prepare emergency preparedness plans. These measures must ensure drinking water security and guarantee, in the event of a crisis, a minimum water supply to meet the basic water requirements. The U.S. Environmental Protection Agency (EPA) developed response protocols to contamination threats and incidents including terrorist acts on drinking water systems. In China, there were 6677 water pollution accidents from 2000 to 2008, many of which threatened the safety of water sources, (Zhang et al. 2010). To address such emergencies the government of China has strengthen its legislation and administration of environmental protection, and enhanced its emergency responses.

Marsily (2002) advocated the idea of "sanctuarizing" large areas where water is the natural element to be protected in order to guarantee long-term water quality, especially for securing drinking water supply. In these zones, called "Natural Hydrological Parks", polluting human activities, mainly agricultural but also industrial, are prohibited or strictly regulated. The idea caught on in France and such zones are being created in many regions. The establishment of Natural Hydrological Parks is fully compatible with the European Water Framework Directive (WFD) which introduced in 2000 the concept of good ecological quality for water bodies. They are also in line with the policy of the protection of "priority catchments" decided in France in 2007, within the framework of the "Grenelle de l'Environnement".

External hard security issues related to water

Externally, water security includes the cross-border dimension. Many surface and groundwater systems are not confined to political boundaries and in consequence, the water availability of one country may depend on the water use (or misuse) of the upstream.

The inventory of international river basins identifies 263 transboundary watercourses covering 145 riparian countries. In these basins, which account for almost half of the surface of the continents and two thirds of the world's watersheds, flows 60% of the world's surface water and live 40% of the world's population. Another significant indicator is the number of countries sharing the same basin, such as the Danube with 17 riparian countries, the Nile, the Rhine, the Congo with more than 9, the Jordan and the Ganges with 5 riparian countries. In spite of these difficulties, the riparian countries in the majority of cases come to organize themselves in order to cooperate to resolve the disputes concerning shared fresh waters. Over the past 60 years, cases of cooperation and agreements have been twice as many as those giving rise to conflict: the database developed over this period (OSU et al. 2002) reports 37 acute conflicts (Violent), against nearly 300 international agreements negotiated or signed.

The UN has played a pivotal role in the promotion of shared water regulations, culminating in the adoption of the 1997 UN Convention on the Law of the Non-Navigational Uses of International Watercourses. It is the only text defining universal principles for the protection and management of transboundary watercourses, principles based on: (i) the concept of equitable and reasonable utilization of the resource, (ii) The obligation not to cause significant damage to other riparian States, and (iii) the general obligation to cooperate. Seventeen years after its adoption by the UN General Assembly, this convention has just come into effect.[1]

Much less known are the international groundwater contained in transboundary aquifer systems, 273 of which were inventoried by UNESCO's WHYMAP Global Program for Hydrogeological Mapping and Evaluation (Jarvis 2006; UNESCO 2011). Initiatives have already made it possible to put in place joint management mechanisms by the riparian countries: a common authority to manage the Nubian Sandstone Aquifer System and a consultation mechanism on the North Sahara Aquifer System, SASS. In Africa, these two examples are still the only cases of international agreements specifically targeting groundwater. Other cases exist elsewhere in the world, notably in Europe, but in very small numbers. In fact, it is in agreements on the use of river basins that groundwater may be mentioned: of the 400 treaties currently in force relating to transboundary freshwater, 100 mention groundwater, only about 10 of which are Specific to the joint management of transboundary aquifers (Matsumuto 2002).

As for international regulation, it has gradually evolved over the last 50 years, from the "Helsinki Rules" in 1966 and the reference to the watershed which may include shallow groundwater, until the adoption by the General Assembly of the

[1]Entry into force 17 August 2014; https://treaties.un.org/pages/ViewDetails.aspx?src=TREATY&mtdsg_no=XXVII-12&chapter=27&clang=_en.

United Nations on December 11, 2008, of the Resolution 63–124 on the draft of the "Law on Transboundary Aquifers", which can serve as a basis for the elaboration of an international convention. The draft was re-examined by the General Assembly of 9 December 2011, which refers the protagonists to the rules of good neighborliness, by "encouraging the States concerned to take appropriate measures and to enter into bilateral agreements for the appropriate management of their transboundary aquifers".

Another dimension of Hard water security concerns the difficulty of supplying virtual water and represents the minimum quantity of basic foodstuffs which must be produced locally to ensure the survival of the population in the event of a serious crisis or major conflict, which would make food import impossible. Water security is then directly linked to food security.

2.3.2 *The National Water Security as a Holistic Approach*

Water needs and the water stress thresholds

Whereas the notion of water stress is naturally connected with the difficulty of supplying population with water resources, the lack of access to safe drinking water at national scale is often determined not by physical water scarcity but by the lack of water infrastructure, the lack of proper technologies and to the lack of sufficient human and material resources. According to Gleik (1996), all water requirements for basic human needs (Basic Water Requirements) account for only relatively small water quantities estimated at 18 m^3/year/capita. However, the specific water consumption depends on the living standards and may thus be largely higher than the Basic Water Requirements but still moderate.

One may admit that Direct water demand (collectivity, industry, tourist), represents relatively low quantity, in any case far below the Falkenmark "severe water stress" threshold of 500 m^3/year/capita. Thus, whatever the water stress situation of a country, the direct water demand supply can be provided through suitable sources provided that the country has the necessary financial and technical means. The agricultural water needs, however, are much higher and the abundance or scarcity has to be appreciated with regard to agricultural uses of water. The Falkenmark thresholds (1000 and 1700 m^3/year/capita) are on such a scale that they should be regarded as representing the water needed for the food production; meaning that food needs, exclusively related to the water availability, are supposed to be provided by irrigation. In many countries, this has led to large water programs in order to eliminate risk and variability and to promote intensive irrigated agriculture so as to meet food security challenge. The central fact is that irrigation consumes a lot of water and its development depends first and foremost on the availability of water and soil resources.

Chahed et al. (2008) noticed that the Falkenmark Water Stress Indicator doesn't take into account the total potential of the water resources and it is not based on the real evaluation of the water demand. It ignores the important part of agriculture production locally produced without recourse to irrigation (rainfed agriculture, pasture) and, for some countries, the no less important part of food products derived from international trade. Moreover, this gives rise to increasing concern to water specialists who fair that the widespread use of this kind of water scarcity indicators risks becoming ‘a fully developed, if misguided, environmental orthodoxy’ (Zeitoun et al. 2013). In a paper by Falkenmark (2013), the water scarcity issues were extended to cover different scales, distinguishing between blue water scarcity (human health, energy supply, industrial production, irrigation, etc.), and green water scarcity (food production, environment). Different country preconditions are identified including countries that lack either green water and blue water to ensure food self-reliance, countries that lack enough green water but have sufficient blue water to compensate by irrigation and countries that have enough green water or/and blue water.

Nevertheless, and despite these new perspectives, the Water Stress Indicators continue to be adopted at different scales including by the specialized international organizations. These criteria applied in a uniform way to all the countries do not consider the disparities which characterize the different kinds of water resources in different regions, the mode of their exploitation and use, the human and material resources employed in their management. Obviously, this appears as an ad hoc qualification of water scarcity situation of a country which is not without effect on the direction of investments and the orientation of technological development, as investments in water systems are often heavy for which international donor assistance plays an important role, especially in developing countries. In addition, it is not easily justifiable to speak about water scarcity without referring to the total water resource, including the pressure-reducing feature of food imports, (Zeitoun et al. 2013).

Securing National water supply by non conventional sources

Strengthening national water security also depends on the country’s capacity to find solutions to develop and increase its potential for meeting water demand. It is possible to subdivide the modes of development and growth of the resource into two main categories; There are, on the one hand, methods that aimed at increasing the potentialities of conventional resources by controlling the use of water and by exploiting resources that escape mobilization by conventional methods (small-scale hydro resources, rain water harvesting, soil water reserve); There are also methods which aim at increasing the availability by producing additional quantities of water obtained by non conventional methods (desalination, waste water reuse).

Non conventional water production techniques can constitute essential components of the freshwater supply system and may be vital in some countries. This is the case for desalination of brackish or sea water, used on a large scale in several countries, in particular in the Middle East region. Endowed with the least fresh water resources around the world The Gulf Cooperation Council (GCC) countries, have largely invested in desalination plants to meet the increasing demand from various sectors (municipality, industry, tourist). In-terms of capacity, the desalination system within the GCC countries represents, in 2013, almost 70% of the total global capacity (Al Hashemi et al. 2014).

Nevertheless, non conventional water production and in particular desalination, are most energy intensive and may be harmful to the environment. Thereby they are costly, both in terms of energy use and in terms of environmental impacts, in particular with regard to greenhouse gas emissions which could exacerbate climate change. While highlighting the invaluable role of desalinated water in fulfilling the demands of the GCC States and their economies, Al-Farra (2014) draws attention to the fact that the dependency of the GCC States on desalination has severely strained their national budgets and caused irreparable damage to their ecosystems.

Despite technological advances that result in a constant decrease of the cost of producing freshwater by desalination, desalination still have high cost (including environmental cost) when compared to other conventional water supplies. The use of desalinated water will remain, for a long time to come, limited only to water uses that are supposed to enhance water uses and justify its production by relatively expensive means. Indeed first experiences in Spain are highlighting that certain important issues related to the agricultural use of seawater desalination can become a barrier to its spread for crop irrigation, (Martínez-Alvarez et al. 2016).

The national water security issue

Considering arid and semi-arid countries, the starting point of national water security is to raise awareness on the limits of the resource. As there are not enough resources to meet all demands, the competition for water becomes more acute and the uses of the same hydrological unit or basin become interdependent. The realization of hydraulic installations to mobilize surface water and groundwater resources (dams, boreholes, reservoirs, transfer chains, etc.) is very efficient in securing water supply and use. It is well perceived, especially in water scarce countries, because it helps cope with scarcity and foster the development of economic activities. However, the management of the resource admits a concrete limit which is the physical potential of the resource.

Within these limits, water resource needs to be developed and managed in such a way as to guarantee the conservation of the natural environments and the protection of the resource. It is necessary, in these conditions to conceive appropriate policy approach to reconcile the various water uses in an equitable and balanced way. That is a very delicate subject because, the measures and trade-offs to implement involve sensitive issues related to a vital resource (legal nature of water, right of access to the resource, economic value of water, geographical unity of the resource, participation, etc.). It also affects the highly water-consuming sector, agriculture, which is the sector of primary economy, intimately connected to the land reflex and the sensitive issues of subsistence, rural development, heritage preservation, food security, social peace and political stability.

Water development programs have to clearly identify all potential sources of risks and to establish policies that ensure long term sustainable water management, tacking into account climatic change impacts. Control of the risks incurred by the resource is a key issue that needs to be recognized as a fundamental element of water resources management. In particular, long term water programs should focus on sustainability.

Furthermore hydraulic structures and water resource management might in themselves have a number of worrying and significant environmental impacts and could create hazards and cause damage. Security arrangements should be considered to prevent against hydraulic disorders, natural disasters or malicious actions, to put in place general planning of interventions and to conceive measures to supply the vital centers of the country in the event of danger.

The water security of a nation should ideally refer to the total sum of the water uses. It is not just about resource mobilization and use, but about a holistic approach to all water resources and their interconnections where one recognizes its multifunctional character, including its environmental function and its socio-economic value. This means also that it is not just about technical and management issues but as well about the political and social side of water security which recovers many facets: ethics, justice, equity, culture. Ultimately, the issue is to put in place a sustainable water policy that improves the governance arrangements toward a global integrative approach to water security. This includes the development of the legislative and institutional framework, the emancipation of technical and human resources and the elaboration of economic and regulatory instruments.

In fine, water security becomes truly meaningful at the national scale only if it benefits all the population and promotes sustainable economic growth, employment and social cohesion. To achieve these objectives, water managers and decision-makers need to be able to understand the direct and indirect relationships between the water sector and the national economy as well as the mechanisms by which the different sectoral policies, even those that are not specific to the water sector, can have implications for resource management.

References

AFD (2015) A question of development, synthesis of AFD studies and research, Facing Water Scarcity in the Mediterranean. http://www.afd.fr/webdav/site/afd/shared/PUBLICATIONS/RECHERCHE/Scientifiques/Question-developpement/23-VA-question-developpement.pdf, consulted on 15 Jan 2017

Al Hashemi R, Zarreen S, Al Raisi A, Al Marzooqi FA, Hasan SW (2014) A review of desalination trends in Gulf cooperation council countries. Int Interdiscip J Sci Res 1:72–96

Al-Farra T (2014) Water security in the Gulf Region. Gulf cooperation council's challenges and prospects, AlJazeera Center for Studies

Asian Development Bank (2016) Asian water development outlook 2016: strengthening water security in Asia and the Pacific. Mandaluyong City, Philippines

Batisse M (2005) Du désert jusqu'à l'eau; 1948–1974; La question de l'eau, de la «Zone aride» à la «Décennie Hydrologique». Cahiers Histoire de l'UNESCO, N°4

Besbes M (1998) Gestion durable des ressources en eau, éléments pour l'élaboration d'une stratégie, *int.rep.* Tunisian Institute for Strategic Studies, Tunis

Besbes M, Chahed J, Bargaoui Z, Mattoussi MS, Mhiri A (2002) Water future, a new challenge for Tunisia, *int.rep.* Tunisian Institute for Strategic Studies, Tunis

Bigas H (2012) The global water crisis: addressing an urgent security issue. Papers for the InterAction Council, 2011–2012. Hamilton, Canada. United Nations University-Institute for Water, Environment and Health

Bigas H (2013) Water security and the global water agenda: a UN-water analytical brief. United Nations University-Institute for Water, Environment and Health
Cárdenas YV (2007) La construction sociale du statut juridique de l'eau en Amérique du nord, *Lex Electronica*, 12(2)
Chahed J, Hamdane A, Besbes M (2008) A comprehensive water balance of Tunisia: blue water, green water and virtual water. Water Int 33(4):415–424
De Marsily G (2002) Vers des parcs naturels hydrologiques. *Le Monde Diplomatique*; *Manière de voir* 65: La Ruée vers l'Eau
DRIEE Ile de France (2014) http://www.driee.ile-de-france.developpement-durable.gouv.fr/l-aquifere-multicouche-de-l-albien-neocomien-a2009.html, consulted on 13 Jan 2017
Falkenmark M (2013) Growing water scarcity in agriculture: future challenge to global water security. Philos Trans Royal Soc London Math Phys Eng Sci 371(2002):20120410
Falkenmark M, Lundquist J, Widstrand C (1989) Macro-scale water scarcity requires micro-scale approaches : aspects of vulnerability in semi-arid development. Nat Resour Forum 13(4):258–267
FAO (1977). Le droit des eaux dans certains pays européens, Section du droit agraire et des eaux, Etude législative N°10, Rome, 1977
FAO (1996) Rome declaration on food security and world food summit plan of action. http://www.fao.org/docrep/003/w3613e/w3613e00.htm, consulted on 15 Jan 2017
FAO (2003) Review of world water resources by country; water report no. 23. FAO: Rome, Italy
FAO-WWC (2015) Toward a water and food secure future, critical perspectives for policy-makers. http://www.fao.org/3/a-i4560e.pdf, consulted on 15 Jan 2017
Fatić A (2002) Conventional and unconventional -hard and soft- security: the distinction. SEER J Lab Soc Aff East Eur 5(3):93–98
Gaonac'h A (1999) La nature juridique de l'eau. Edition Johanet, France
Gerlak AK, Mukhtarov F (2015) Ways of knowing'water: integrated water resources management and water security as complementary discourses. Int Environ Agreem Polit Law Econ 15(3):257–272
Gleik PH (1996) Basic requirement for human activities. Meet Basic Needs Water Int 21:83–92
Global Water Partnership (2010) Water security for development: insights from African partnerships in action. GWP, Stockholm, Sweden
Grey D, Sadoff CW (2007) Sink or swim? Water security for growth and development. Water policy 9(6):545–571
Hoekstra AY, Mekonnen MM (2012) The water footprint of humanity. Proc Natl Acad Sci 109(9):3232–3237
Houdret A, Kramer A, Carius A (2010) The water security nexus: challenges and opportunities for development cooperation. *Technische Zusammenarbeit (GTZ) GmbH*, International Water Policy and Infrastructure Programme
ITES (2011) Etude stratégique - Eau 2050, Annexe 1, le Bilan Hydrique National, *int.rep.* Tunisian Institute for Strategic Studies, Tunis
Jarvis WT (2006) Transboundary groundwater, geopolitical consequences, commons sense, and the law of the hidden sea, Ph.D., Oregon State University
Lankford B, Bakker K, Zeitoun M, Conway D (2013) Water security: principles, perspectives and practices. Routledge
Martínez-Alvarez V, Martin-Gorriz B, Soto-García M (2016) Seawater desalination for crop irrigation—a review of current experiences and revealed key issues. Desalination 381:58–70
Matsumoto K (2002) Transboundary groundwater and international law: past practices and current implications. MSc, Dep. Geosciences, Oregon State University
OSU (Oregon State University) UNEP and FAO (2002) Atlas of International Freshwater Agreements
Pahl-Wostl C, Bhaduri A, Gupta J (2016) Handbook on water security, Edward Elgar Publishing, UK

Paquerot S (2004) De l'intuition à la précision conceptuelle, les différentes perceptions du/des bien/s public/s à l'échelle mondiale : l'exemple de l'eau douce, «Global Flows, Dissent and Diversity: the New Agenda», Montréal - May 14–15, 2004

Pozzo B (2000) L'eau et les droits de propriété en droit civil et en common law, in *Droit de propriété, économie et environnement*, M.Falque M.Massenet Ed. Dalloz

Raskin P, Gleick PH, Kirshen P, Pontius RG, Strzepek K (1997) Water futures, assessment of long-range patterns and problems. Stockholm Environment Institute, for the United Nations

Sadoff CW, Hall JW, Grey D, Aerts JCJH, Ait-Kadi M, Brown C, Cox A, Dadson S, Garrick D, Kelman J, McCornick P, Ringler C, Rosegrant M, Whittington D, Wiberg D (2015) Securing water, sustaining growth: report of the GWP/OECD task force on water security and sustainable growth. University of Oxford, UK

Seckler D, Barker R, Amarasinghe U (1999) Water scarcity in the twenty-first century. Int J Water Resour Dev 15(1–2):29–42

Sullivan C (2002) Calculating a water poverty index. World Dev 30(7):1195–1210

Sullivan CA, Meigh JR, Giacomello AM (2003) The water poverty index: development and application at the community scale. In Natural Resources Forum 27(3):189–199, Blackwell Publishing Ltd

UNESCO (2011). http://www.unesco.org/water/news/aquiferes_transfrontaliers.shtml, consulted on 15 jan 2017

US EPA. (2011). http://water.epa.gov/infrastructure/watersecurity/lawsregs/indexcfm#bioterror, consulted on 15 Jan 2017

World Bank (2005) Water resources, growth and development, a working paper for discussion prepared by The World Bank for the Panel of Finance Ministers The U.N. Commission on Sustainable Development, 18 April 2005

World Water Council (2015) Water: Fit to finance? Catalyzing national growth through investment in water security, Report of the High Level Panel on Financing Infrastructure for a Water-Secure World April 2015, Marseille, France

Zeitoun M (2011) The global web of national water security. Glob Policy 2:286–296

Zeitoun M, Lankford B, Bakker K, Conway D (2013) A battle of ideas for water security. In: Lankford B, Bakker K, Zeitoun M, Conway D (eds) Water security: principles, perspectives, practice. Routledge, London

Zhang XJ, Chen C, Lin PF, Hou AX, Niu ZB, Wang J (2010) Emergency drinking water treatment during source water pollution accidents in China: origin analysis, framework and technologies. Environ Sci Technol 45(1):161–167

Zimmer D, Renault D (2003) Virtual water in food production and global trade: review of methodological issues and preliminary results. Virtual Water Trade Proc Int Expert Meet Virtual Water Trade, Value of Water Res Rep Series 12(1):1–19

Chapter 3
Fifty Years of Water Policies, 1960–2010

Contents

M. Besbes et al., *National Water Security*,
https://doi.org/10.1007/978-3-319-75499-4_3

3.1 Introduction

After concretization of three major water Master plans, and eleven five-years economic and social development plans over the past fifty years, the country has implemented a network of dams, transfer facilities and water wells, allowing to ensure relative safety about drinking water supply of 100% of the urban population and 94% of the rural areas. Building on water infrastructure, irrigation has been developed on nearly 420,000 ha, or 8% of arable land which contribute nearly a third of the agricultural production value. These water facilities are involved, in addition, to mitigate the flood-related damage and droughts. Urban sanitation has also developed favorably in large cities. All these achievements contribute to improve food and health conditions of the Tunisian people.

However, any development of water resources policy admits an objective limit which is the threshold of the resource. With completion of most hydraulic mobilization facilities, the policy of increasing supply of conventional resources is beginning to reach its limits. First fruits of meeting these limits, and corollaries of accelerated water resources development were clearly identified since the early 1990s: overexploitation of many shallow and deep aquifers, water salinization in some of them, soil salinization in some irrigated areas, development of drinking water and irrigation at the expense of groundwater recharge, reduced river transport capacity downstream of large dams, silting and gradual decrease of reservoir dams capacities.

In this context, integrated management of the resource, including demand management, is gradually introduced to meet the growing needs expressed by various economic sectors. From a policy based on water supply, strategies evolve towards a policy that aims to improve efficiency of hydraulic systems and control of water demand. Thus, Tunisia has been engaged in a new phase of managing its water resources. Important changes are taking place and reforms are implemented to protect water resources, control production and use of non conventional water resources, better manage hydrological risks, anticipate conflicting sectoral interests by mobilizing all stakeholders, modernize the regulatory framework and develop appropriate planning.

3.2 The Geographical Framework

The physical context and relief

Located at the Northern tip of Africa and extended over seven degrees of latitude, Tunisia offers a succession of hydro-climatic types: (a) sub-humid in the North, (b) arid in central Tunisia, (c) hyper arid and desert for the entire South. These climatic types articulate with the four major geographical provinces:

(i) In the North West, the reliefs of Tell and Dorsal oriented SW-NE between 500 and 1500 m, frame the most fertile plains of the country. Orthogonal to the direction of the prevailing winds from the NW, orientation reliefs promotes significant rainfall, which makes this area as the wettest of the country: the main river, the Medjerda Oued, fully controlled by a series of large storage dams, crosses the region from west to east to unclog the North of the Tunis Gulf (Fig. 3.1).

(ii) In the Central West, the steppe land extends from the massive mountainous and arid high plateaus of Gafsa, alternating with large alluvial plains richly equipped with groundwater, closely linked to the system of wadis Merguellil, Zeroud and Bayech.

(iii) In East Tunisia, the coastal plains region includes the Northern Sahel and Sahel of Sfax with poor water resources, and the plain of Gabes-Medenine deposit bearing the Djeffara aquifer system.

(iii) The South in the Saharan Platform, marked by the immensity of Chott Djerid basin and dunes of the Grand Erg Oriental, connects with Djeffara coastal plain to the east by the Dahar plateau. This is the area of large Saharan aquifers of the Continental Intercalary and Terminal Complex basins.

The drainage hydrographic system

North of the Tunisian dorsal, the hydrographic system is fairly well organized and hierarchically structured in three watersheds (Fig. 3.1): (i) the far North-Ichkeul basin, (ii) that of Medjerda whose upstream third is in Algeria, (iii) the Miliane basin and Cape Bon wadis. Because of its climate, landscape and rainfall, North of Tunisia is the only region able to maintain exoreic perennial river flows, and provides 82% of total surface water supplies of the country.

South of the dorsal, arid climate, degraded vegetation and intense evaporation do not allow the formation of river systems as well organized, and the concept of river basin often loses its relevance; the flow structure is generally a degraded network except in the upstream basins: outside the strictly coastal basins, wadis are no longer able to make their way to the sea and water infiltrates in the crossed permeable alluvial plains then evaporates in large inland depressions and Chotts Sebkhas. It is the privileged area of large endorheic systems: (i) The Sebkha Kelbia basin with fully artificialized wadis Nebhana, Merguellil and Zeroud, (ii) The Chott El Gharsa-Sebkhet Nouail basin, (iii) The Chott Jerid basin that drains Saharan platform, (iv) the Sahel of Sousse and Sfax coastal basins, (v) The catchment area of the Djeffara coastal plain.

The Oued Medjerda forms the central artery of the Tunisian water system. With a catchment area of 23,700 km^2, among which 16,100 in Tunisia, Medjerda has ten tributaries, including 5 on the right bank (Mellègue, Tessa, Siliana, Khalled, Chafrou) and 5 on the left bank (Rarai, Bou Heurtma, Kasseb, Beja, Zerga).

Basins of the extreme North (Oued Barbara, Melilla, Kebir Zouara) and Ichkeul (Joumine, Rhezala, Sejenane, Douimis) extend over a total area of 5000 km^2 and have a remarkable intake, due to a well-stocked rainfall.

Fig. 3.1 General situation map and watersheds (According to the World Water Assessment Programme WWAP 2009)

Brief geological framework, and aquifers occurrence

The geological history has endowed Tunisia with rich aquifer deposits, of which the vast Saharan basin containing considerable reserves, a multitude of aquifers with reduced dimensions punctuating the North, and aquifer basins of relatively large dimensions at the Centre.

Structurally, Tunisia is clearly divided into a Northern part, subject to the Alpine tectonics, where the geological strata are strongly folded and conducive to the accumulation of medium-size aquifers, and a southern part characterized by tectonics platform, which dominates the large sedimentary basin of Northern Sahara. Tunisia is relatively poor in surface waters and aquifers there have always been the main water resource. The apparent paradox of Tunisia's regional water resources specificity is that surface water is almost exclusively located in the North, while groundwater is located in central and southern regions. This distribution accurately combines geol ogy with rainfall: the far North is poor in aquifers, despite higher rainfall (reaching more than 1000 mm/year), but due to a basement where predominates the Numidian flysch low permeability. In Northern Tunisia, where geological layers are tectonized and creased, there are many small-scale layers, which can grow in the plains where groundwater is exclusively related to alluvial sediments such as those of Ghardimaou or Mornag.

The geological structure of the central Tunisia is in turn dominated by the Saharan Atlas, crossing the country in a SW-NE diagonal, the Tunisian Ridge. This structure is affected by many collapse pits, seat of important aquifers deposits reaching hundreds of meters thick: Gafsa, Kasserine, Sidi Bouzid, Kairouan, and Grombalia. Southern Tunisia, largely desert, is dominated by the Saharan platform where major aquifers lie hundreds of thousands km^2 shared by Algeria, Libya and Tunisia. For the case of Continental Intercalary and Terminal Complex layers, their gigantic water reserves are weakly renewable and only partially usable.

3.3 Water Management, A Long Tradition in Tunisia

3.3.1 A Brief History of Water Management in Tunisia

Water development traditions were initiated in Tunisia by Berber populations many times before the Phoenician period. Then Carthage became the second most important city in the Roman Empire, and the construction of the Carthage aqueduct, about 132 km, dates back to the middle of the 2nd century: it must capture Zaghouan springs to meet the growing water demand in the highly urbanized city. Roman hydraulic works in Tunisia (Gauckler 1897) cover both use of local resources through rainwater harvesting infrastructures, and large-scale water transfers to major water consumption regions. These two aspects on water resources development came to dominate the water management modes throughout the history of Tunisia.

From the seventh century, the Arab-Muslim contribution was marked by a new legal and institutional approach to water management, the rehabilitation of old facilities and expansion of existing hydraulic techniques to new areas of the country. The Aghlabids dynasty furnished an important conservation effort of facilities and organized the supply of Kairouan by the Cherichira aqueduct, along 35 km and the Aghlabid basins where the Wadi Merguellil floods were decanted and stored. As for the Roman aqueduct of Zaghouan-Carthage, it crossed many vicissitudes before being restored by the Hafsids under the reign of al-Mustansir (1249–1277). The work focused on the restoration of 116 km of the old aqueduct and adding a monumental facility leading water to Tunis (Solignac 1936). The maintenance of the aqueduct was abandoned after the Hafsids, and it was only in 1852 that Sadok Bey ordered again its restoration to improve water supply of Tunis, for centuries before reduced to water cisterns use.

In irrigation field, the desert south knew an effective water management period with the development of water organized societies in oases: rigorous water distribution in the Tozeur oasis by Ibn Chabbat created since the thirteenth century a new water distribution model by fixed rotation to users. Despite a rather loose agrarian presence in the country and the predominance of pastoralism, several flourishing hydraulic schemes grew, using flood irrigation, especially in the high steppe as in the plain of Kairouan. In the Sahel on the east coast, collecting runoff from local catchments has maintained over centuries large areas of olive groves (El Amami 1984).

Technological progress was introduced in Tunisia with various waves of Andalusians immigration. The waterwheel, improved version of the Persian wheel, was the first attempt of bailing mechanization and the basis of all irrigated agriculture in the North and coastal areas. The water catchment by underground galleries (foggaras) allowed supply of drinking water to towns and oases irrigation. In 1622, the Andalusians have received, with the support of Youssef Dey, permission to build a dam with locks and bridge on the Medjerda River at El Bathan near Tunis. This work, with two main pipelines supplying an irrigated olive grove area (2500 ha), is the first major irrigation scheme in the modern Tunisian history.

3.3.2 Birth of Tunisian Modern Hydrology

First priorities of colonial administration included the knowledge of natural resources, mainly water. After the french occupation, the Scientific Mission of Tunisia in 1884–1887 (Morin and Memmi 1972), brought relatively complete geological knowledge about the Center, South, North and North East. This work served to prepare the provisional Geological Map of Tunisia at the 1/800,000 scale (Aubert 1892). On this same scale was also published the provisional soils map of Tunisia by Agavonoff and Yankovitch. The final results of the Scientific Mission were published between 1907 and 1913 (Thomas and Pervinquière 1907, 1908, 1913). This work had an extension with the Geological Survey of Central Tunisia and a map at the

1/200,000 scale, always present. Later, was published by M. Solignac the imposing "Geological Survey of Northern Tunisia", and in 1931 the Geological Map of Tunisia 1/500,000 scale in 3 sheets. Following theses achievements, a new geological map at 1/500,000, by G. Castany, was published in 1952.

With geological mapping, drilling played a key role in aquifers recognition programs. By 1881, artesian wells around Tunisian chotts were drilled on behalf of the Inland Sea project (Burollet 1995) but the first listed water well (Hamza 1992) was drilled in 1886 in Sfax by the Mining Service, it was 150 m deep and wore No. 1 of the Water Resources Inventory. Drilling No. 2 was completed in 1888 and was 334 m deep. In 1991–93, drillings in Djerba confirmed the continuity of the Miocene sands aquifer between this island and the continent. Recognition by drilling of all south aquifers then underwent intense development.

The first comprehensive assessment of water resources of the country is due to Boulle, a public works engineer, in his report "The development of the Regency waters" dated from 1896. He already reported 3000 inventoried water points and gave a summary overview of possible aquifers. Meanwhile, conceptual studies were underway and in 1910 were presented several notes on the Tunisian hydrogeology at the French Academy of Sciences and the French Geological Society. The Tunisian administration instructed the Public Works Department (DTP) to establish a systematic inventory of water resources of the country, a task that required close coordination of the various concerned departments: Meteorological Service, Mining Service, Geological Service, Hydraulics Service, and Topographical Service. The director of the hydraulics department, M. Gosselin, and his successor J. Tixeront, were responsible for this coordination and set up a special organization called the Office of Water Resources Inventory: the BIRH, entrusted to E. Berkaloff. All modern scientific methods at that time were taken to carry out the mission of BIRH (Gosselin 1969). The first "Catalogue of Waters" has been prepared which ultimately in 1950 registered 30,000 water points. Investigations within the BIRH activities have traditionally focused on climatology and rainfall, surface water and groundwater. Regarding groundwater, the BIRH used the detailed geological surveys completed by observation and aerial photography, by geophysical prospecting and electrical logging. Chemical composition has been developed and proved to be a valuable mean of water investigation. Several artesian wells were drilled, the observance of which has made it possible to develop practical methods for determining underground hydraulic characteristics, (Gosselin 1952).

During the Second World War, the BIRH was instructed to prepare a large number of water points in pre-Saharan regions. With the assistance of hydro geologists from Algeria and Morocco, it was possible to make an inventory of water resources of the entire Saharan territory and resolve problems raised: the documentation collected allowed to publish, after the war, a seven sheets hydrogeological map at the 1/200,000 scale. Geologists from Algeria and Morocco, who had worked with Tunisian methods, returned to their respective countries and tried to organize the establishment of similar inventories (Gosselin 1969). A few years later, with the independence of Morocco and Tunisia in 1956, these methods were introduced in France at BRGM, the French Office of Geological and Mining Research.

The Tunisian Meteorological Service was established in 1885, and in a few years it had implemented an important observation network (Bargaoui 1993). The service was conducted in 1894–1930 by G. Ginestous, author in 1906 of the "Study on the climate of Tunisia." The wealth of rich information provided by the network between 1900 and 1940 authorized the publication, by H. Gaussen and A. Vernet in 1952, of the Precipitation Map of Tunisia at the 1/500,000 scale, a map currently in force. Regarding surface water, measures of wadi flows by the BIRH teams began in 1925 on the Medjerda, the Kebir and then spread to all streams.

As for the Geological Survey under the direction of J. Archambault in 1945, it occupied a privileged position in the groundwater recognition program, thanks to the expertise of confirmed geologists including A. Azzouz, G. Castany, R. Degallier and H. Maazoun. All of rainfall and hydrological data collected allowed the publication of the first water balance of Tunisia (Tixeront and Berkaloff 1954), which referred to rainwater resources equivalent to 33 km^3/year, and runoff of about 2 km^3/year: these magnitudes have retained all their validity until today.

3.3.3 Hydraulics and Irrigation Development

As soon as the Protectorate was established in 1881, an exploration policy for aquifers by drilling was implemented, which will develop in the south and center of the country to expand palm groves, arboriculture and forage crops. At the same time, irrigated areas by shallow wells began to expand, particularly in the Tunis region, and in the Cape Bon specialized in vegetable gardens and citrus fruits, causing groundwater resources overexploitation and already questioning their conservation.

The large irrigation developments

At the end of the 30s major agricultural drainage and flood protection projects were undertaken, among which completion of the main channels for the Lower Valley of Medjerda irrigation and drainage project.

Before the Second World War, Tunisia had not yet engaged major hydraulic works and a single large dam had been built on the Kebir River, in 1928, for drinking water supply of the capital Tunis. After the war, the policy of large dams found a favorable field in the Medjerda basin, where land is extensive and fertile, and where irrigation could effectively alleviate the uneven rainfall. Such options will reinforce the emergence of the regional hydro-agricultural development plans:

(i) In the Northern region: irrigation possibilities of Medjerda Basin were estimated at 100,000 ha, achievable through construction of several large dams. In the vicinity of Tunis, irrigation of specific areas, covering up to 10,000 ha were planned by reuse of treated wastewater from Charguia station. In the Northern region, large dams totaling a mobilization capacity of 400 million m^3/year, and irrigated perimeter of Lower Medjerda valley (40,000 ha) have concretized the largest hydro-agricultural project of the Protectorate Administration. This program, which is part of the "Equipment and Modernization Plan of Tunisia" (Chevalier 1950) and which has not been completed by the colonial administration, will be pursued by the Tunisian government after independence and formed the framework on which was developed the Medjerda River and its tributaries Master plan.

(ii) In the Central Region: the wadis are too irregular to consider solutions such as storage dams. However, groundwater that have been recognized and partly captured, was used to establish an irrigation program on 11,000 ha. The flood water spreading irrigation has been practiced here since a long time, particularly well controlled in Kairouan plain: modernization works had been carried out to better control siltation and flood water distribution mechanisms.

(iii) In the South Region: major projects included safeguarding the ancient oases and creating new ones in the Djerid and Nefzaoua. In Gabes-Djeffara, were carried out works of flood water spreading and retention in the wadis of the Dahar.

The public water supply

Between 1880 and 1914, the water supply of Tunis being temporarily assured, the effort was focused on other large cities: (i) in 1905, Sousse in the Sahel is supplied from the Bou Hafna draining galleries by means of a 125 km pipe at a flow rate of 2500 m^3/day; (ii) in 1914, Sfax in the South was fed by a cast iron pipe on 175 km which brings water from Sbeïtla springs at a flow rate of 8000 m^3/day.

Feeding Tunis was improved by connection of new supplies to the Zaghouan aqueduct, and since 1911 a first disinfection process with bleach is installed. After 1914, drought and cities consumption rise resulted in severe water shortages for Tunis, Sousse and Bizerte, requiring new supply systems. In Tunis, the rate increased from 6000 to 70,000 m^3/day, an increase made possible by commissioning the dam of O. Kébir and abstracting groundwater from Khledia, Jouggar and Manouba aquifers. The participation of private sector in the management of public services was common: most cities conceded the management of their drinking water systems to private companies.

Urban sanitation

The first rational sanitation study of Tunis was carried out in 1889, to provide the city by a modern sewerage system. The works, pipe laying and construction of a pumping station, began in 1891, and similar works were undertaken for other major cities. So that by 1945, these cities had modern sewer systems. After the Second World War, population growth and development of water supply led to an extensive program to strengthen urban sanitation: (i) creation of sewage systems in cities and suburbs hitherto not equipped; (ii) rehabilitation of existing networks; (iii) biological purification of water from the city of Tunis with prospects for the recycling of treated wastewater and sludge. This program included studies of beaches bacteriological contamination in coastal cities and best discharge points. The same program continued after Independence under supervision of municipalities, until the creation of ONAS in 1974.

The water law evolutions

A water legislation was set-up in order to support hydraulic achievements and modernize the traditional water management legislation: (i) The Beylical decree of 24 September 1885 incorporated the public domain watercourses, springs, aqueducts, wells, irrigation channels and canals run by the State in public utility purposes; (ii) The 1897 decree recognized "water users' associations"; (iii) The 1920 decree regulated the conservation and use of Public Domain waters; indeed this was the first Tunisian Water Code, which created the "Special Associations of water Interest"; these associations had to be created at the initiative of water users and put under the authority; (iv) The 1933 Beylical decree integrated groundwater into the public domain, completing thus the 1885 Decree and imposing state ownership of water; property rights were consequently transformed into non-transferable water rights of use.

3.4 The Broad Water Policies

3.4.1 Genesis of the Tunisian Water Project

At Independence (1956), the Medjerda Valley in the North was subject to extensive works within the framework of Tunisia's modernization and equipment plan: construction of dams, irrigation canals, hydroelectric stations, sewerage infrastructures, and soils and water conservation. The Lower Medjerda Valley development project then attracted major interest from the new Tunisian government, which mobilized all national and international resources to ensure the continuity of achievements. The Office for the Development of the Medjerda Valley (OMVVM) was created in 1958 to operate the hydraulic structures, carry out studies and technical achievements, advise and assist farmers in irrigation development.

The "Ten-year Development Perspectives 1962–1971" (State Secretariat for Planning and Finance, 1961) presented the first economic and social development doctrine of independent Tunisia, based on the founding principles of "decolonization, tunisification of production equipments, human promotion, structural reforms and self development". With regard to water resources, these principles were defined in terms of specific targets: "Water scarcity and irregular rainfall oblige Tunisia to use all means to maximize its surface water retention and groundwater exploitation". Actions to be taken included: (i) studies and research for a better knowledge of water and soil resources; (ii) urban water infrastructures for water supply, urban sanitation, and flood control; (iii) agricultural water works, by construction and equipment of wells and boreholes, creating four large dams (Kasseb, Bou Heurtma, Joumine and Nebhana), small dams, and planning for the irrigation of 60,000 ha.

By the early seventies, the main idea underlying agricultural policy was that the extent of the food issue could be reduced by planning to achieve a greater degree of self-sufficiency. This concept placed the food issue at the heart of the State responsibility. With aridity, agricultural intensification through irrigation was considered a salutary solution: conductive to increasing agricultural production, it should provide answers to food questions and contribute to the promotion of rural areas. Tunisia has set up a large-scale hydro-agricultural planning to achieve the political objective of food self-sufficiency; especially as the water project started since colonial times had prepared the ground.

While maintaining a strong public sector, the government has encouraged the private sector which benefited from the administration technical supervision and support of the banking system to set-up modern agriculture. In order to support this agricultural development effort, the State should boost the rate of water resources mobilization. The priority of large scale hydraulics prevailed for the development of the Northern region, and the preference for great projects will mark the technical choices of Tunisia in this region. To achieve its water project, the government had to rely on international cooperation. The water project also needed technical expertise: the first generation of Tunisian engineers followed the same way that pioneers of modern Tunisian hydraulics, and took over by continuing to benefit from technical and financial support from friendly countries, international funding agencies, and consulting firms and contractors services. The Tunisian administration will not only succeed in assimilating the water project of the colonial era and continuing it; it went even further in maximum water resources mobilization by large water infrastructure to promote a modern agricultural sector.

3.4.2 The Tunisian Water Story

The originality of the Tunisian approach was undoubtedly this early perception of the major water issues in relation to the spatial distribution of the resource versus needs: the hydraulic option has been maintained and strengthened by a policy of massive water transfers intended to "correct" water interregional inequalities. These transfers to supply cities with drinking water were common practice. What was new however is that after Independence, water transfers objectives have been widened to cover agricultural water needs. This orientation was clearly expressed at the time of building the Nebhana dam in the sixties. A 130 km line carried water from the dam to eastern coastal region to cover needs of irrigated farms. Developing the maximum amount of surface water through major water infrastructures, transferring water to deficit areas and achieve a real national interconnection, these were the elements of Tunisia's water policy. This policy has helped since 1970s to realize vast water infrastructure programs organized around three Regional Water Master Plans developed respectively for the North and Extreme North (PDEN), the Centre (PDEC) and the South (PDES).

The development of water resources in the Medjerda basin under PDEN was totally in line with this ultimate development of water resources perspective. The decision to build the Sidi Salem Dam (the largest dam of the country) on the Medjerda River, especially to ensure the transfer of water needed to supply drinking water to the cities of Tunis region and the Sahel and to support irrigation in the citrus growing areas of Cape Bon, marked a decisive step in this resolute policy of mass mobilization and transfer of water resources (Zamiti 1985). The Tunisian Administration will make from Sidi Salem dam, flooded in 1982, the centerpiece of the PDEN system. His reservoir, the largest in Tunisia, which had a capacity of 550 million m^3, was carried out to 720 million m^3 in 2000.

In the arid central Tunisia, the agricultural sector had, since Independence, a major political and social role: in addition to its contribution to the national food production effort, the agricultural project objective was the settlement of nomadic pastoralists in the steppe. To achieve this objective, the implemented strategy has targeted the extension of cultivated areas and mobilization of water and soil resources to promote irrigation. Several public irrigated areas have been created in the regions of Kairouan, Sidi Bouzid, Kasserine, and Gafsa.

The implementation of hydro agricultural policy was facilitated by the institutional set up in the early decades of Independence. Thus the Regional Irrigation Development Offices (OMVPI) have played a key role in the completion of management plans. Thirteen regional offices were created between 1958 and 1984, and invested in various missions: (i) creation of new public irrigated areas; (ii) operation and maintenance of water systems; (iii) supervision of irrigators and support for agricultural land development operations (farm credit, inputs supply, collection, marketing and processing of agricultural products). OMVPI were integrated to Regional Commissions for Agricultural Development (CRDA) in 1989.

The Water Code promulgation in 1975 (JORT 1975) has strengthened the central public authority control on water. It helped to implement the water production and supply policy at the national level, and to provide the general vision necessary for the coordination of the technical and financial resources needed to achieve significant improvements of water mobilization, storage and transfer. The Water Code has consecrated the principle of maximum valorization of water on a national scale. This principle, fundamental to the conception of the resource management, constitutes the legal foundation of water policy for interregional transfer and inter sectoral water allocation.

In general, and because of the unequal distribution of resources, Tunisia has opted for expensive hydraulic solutions to lead supply management policy at the national level: water mobilization and transfer projects have been fully realized on state funds, and private initiatives to mobilize water were encouraged with financial subsidies to boreholes digging, construction and equipping of individual irrigation schemes. The Tunisian water program has thus captured a significant share of public investment: as the sixties, budgetary allocations for water supply were around 40% of total credits granted to the agricultural sector.

This water policy, which was clearly focused on increasing supply, was inevitably come up against the limits of the resource, and the first manifestations of reaching these limits were identified in most parts of the country since the early 1990s. Paradoxically, but in a maximum water security logic, it is precisely at this time of awareness on water scarcity and the possible occurrence of future shortages, that will launch the National Strategy of Water Resources Mobilization.

3.4.3 The Necessary Change in Management Processes

Water withdrawals are approaching the order of magnitude of the resource's potential, and the prospect of stabilizing supply arises naturally the resource allocation issue. In these conditions, water reallocations to meet drinking water demand and ensure supply of priority sectors (industry, tourism) can only be done to the detriment of the agricultural sector. The prospect of decreasing good quality of water for agriculture is a key issue in the water problem: it is in total contradiction with the irrigation policy of the last fifty years. This critical question had largely determined the general framework of the Water Sector Study launched by the Ministry of Agriculture in the 1990s (DGRE 1999).

The Water Sector Study diagnosis clearly indicates that water policy, based on supply management, must give way to a demand-side approach aimed at reducing consumption, based on an integrated water resource planning system. The Study, which formalized the awareness of the need for a move towards Integrated Water Resources Management (IWRM), proposes a comprehensive approach whose main ideas are: (i) strengthening monitoring, control, and protection of resources; (ii) developing efficient uses by controlling demand and better valuing uses; (iii) promoting decentralization by strengthening the role of users and the involvement of private operators; (iv) recognizing the economic, social, and environmental value of water; (v) controlling risks. As part of these general principles, the Water Sector Study proposes short term and long term draft solutions.

The short-term solutions are of an operational nature and relate to the technical, financial and organizational aspects that aimed at implementing a water demand management policy: protecting resources, reducing losses, improving efficiency of uses, considering the economic value of the resource, decentralization, and participatory management. Centralized management of water resources lasted until the end of the 1980s. The Ministry of Agriculture services restructuring in 1989 engaged the decentralization of agricultural administration and strengthened the role of Regional Commissions for Agricultural Development (CRDAs). Within the governorate (see Fig. 3.2), CRDAs coordinate the government agricultural policy at the regional level. The CRDAs are responsible for managing great irrigation schemes and help to provide water to users. They rely in this mission on user associations which support the infrastructure operation and distribution of water for irrigation and drinking in rural areas.

3.5 The Main Leverages: Planning and Investment

The Tunisian water program was organized around regional master plans to support the government action in different areas and support economic and social development. Water resources planning has thus become part of an overall long-term vision, in terms of natural regions and across the country. It embodies the major choices of

Fig. 3.2 The governorates map of Tunisia

the Tunisian water policy. Because of the unequal resources-distribution among the different regions, the drinking water transfer to urban centers has emerged very early as a fundamental option for water supply policy. Planning resulted in significant investment strategies whose implementation relied on public, private and international funding opportunities.

3.5.1 The General Framework for Water Planning in Tunisia

Going beyond the sectoral or project-based approach that prevailed during the first half of the 20th Century, Tunisia has opted since the 1960s for water planning to avoid uncoordinated resources development and to ensure balance between needs and water availability. Water planning is an important component of economic and social development plans, and contributes decisively to the achievement of regional and national goals for access to drinking water, irrigated agricultural production to meet food security, and development of disadvantaged regions.

Water planning has two complementary components: (i) the first is relevant to the formulation, over the long term and at the country level, of a comprehensive vision, a number of strategies and orientations to draw the best use of available water resources, and to resolve possible conflicts between competing uses. This type of planning is flexible and may be adapted to modifications and revisions deriving from experience; (ii) the second part is relative to the programming of water projects which permits, in the medium term, to identify priorities and act on investment in the water sector according to the long-term planning.

Long term planning: the regional water planning

To achieve the most efficient water resource management of the country, it was necessary to have coherent plans, valid in the long term. To this end, three master plans of water use covering the three main geographical regions (North, Centre, South) have been developed to better reflect the complexity related to the potential variability of the sub regions resources, the acute problems of water quality, and to the remoteness of consumption centers located mainly in coastal areas compared to production sites.

Medium term planning: the economic and social development plans

General planning is an ongoing process in Tunisia. The first planning act dates back to 1962 with the development of the "Decennial Perspectives" and the first three-year plan 1962–1964. Since, twelve plans for economic and social development, five-year plans since 1972, have been drawn up. The Economic and Social Development Plan is designed as a general policy instrument of the country's development policy and as part of program making, including those relating to the water sector. This process has timidly evolved over the 1990s from a centralized approach to a participatory and decentralized approach, involving the key water stakeholders and taking into account local and regional concerns. Indeed, apart from the great hydraulic works whose planning and implementation are decided at the central level, the various water projects and programs are proposed at the level of regional structures, in this case by the CRDAs and public water agencies (SONEDE, ONAS). The central structures of the Ministry of Agriculture are empowered to monitor the priorities based on government guidelines and to harmonize the programs on the regional and national levels according to available financial resources. In this context, consultations of direct beneficiaries of water projects and programs are carried out after definitive enrollment in the Plan.

The national water mobilization strategies

Since 1990, two successive ten-year water mobilization strategies [Strategy I (1990–2000), Strategy II (2002–2011)] have been developed, following completion of Regional Master Plans. These strategies were essentially physical timetables for water projects, in a supply management view: large dams, hill dams and lakes, boreholes for groundwater recognition and monitoring, spreading facilities and artificial recharge, water treatment plants, etc. Based on a set of actions leading to the total mobilization of groundwater and surface water, these strategies served actually as a support for investment programming in water infrastructures, executed in accordance with the five-year plans development objectives.

- *The first strategy 1990–2000* (DGRE 1990): The 1990–2000 decade was marked by establishment and implementation of the strategy aimed to: (i) the maximum mobilization of available water resources into the limits of acceptable levels of quality and costs; (ii) the soil and water conservation. Starting from a situation where 62% of surface water and 80% of groundwater were mobilized, the strategy initially proposed to "mobilize all the water resources of the country … to achieve by 2000 the provision of 100% of the identified water potential". These objectives were subsequently adjusted to target a 90% mobilization rate with water infrastructure to mobilize and control an additional volume of 1400 million m^3/year for a total cost of 1 900 million Dinars.[1]
- *The second strategy (2002–2011)*: this new program had to continue efforts to mobilize up to 95% of potential resources through implementation of 11 new large dams and 50 hillside dams. Along with supply increases through mobilization, this strategy deliberately opted for the introduction of new integrated resource management objectives, for water savings and demand management, non-conventional water resource development, and water resources protection.

Programs and specific projects to the medium term

While continuing new water resources mobilization as part of management plans and decennial strategies, water sector reforms have been initiated, and specific programs to stimulate a demand management policy have been introduced. The visibility of the water policy was successively marked by: (i) the establishment of political and institutional guidelines for the implementation of the Agricultural Structural Adjustment Plan (1986–1994), which marked the progressive disengagement of the State (agricultural liberalization, reduction of subsidies granted by the State, adjustment of water prices, loss of monopoly for the public offices trade); (ii) the large water programs initiated over the last two decades: The Investment Project in Agricultural Sector (1991–2000); the Investment Project in the Water Sector [PISEAU-1&2 (2001–2007) and (2009–2014)].

[1]On 26/04/2017, the Tunisian Dinar was worth 0.38 € or 0.41 US $.

This policy is characterized by a number of significant and sustainable actions, including: (i) introduction of pricing systems adapted to drinking water and irrigation sectors, in order to improve costs recovery, guiding the irrigated production policy and encouraging the economic valuation of available resource; (ii) development of water-saving programs in various sectors; (iii) encouraging active participation of users in irrigation and drinking water in rural areas through water users associations (Agricultural Development Groups: GDAs); (iv) valuation of non conventional water resources by reusing treated wastewater for agriculture and recreation, and desalination of brackish water for pressing drinking water needs; (v) protection against water pollution, mainly due to urban sanitation programs.

Legal and institutional inadequacies related to planning

The important hydraulic effort and water management, particularly towards achieving mobilization and regulation facilities, and towards extending interconnection and supply networks, must not evade inadequate conceptual and regulatory framework in terms of policies and resource planning. The centralized and binding planning model, based on growing supply in scarce water regions, is becoming increasingly controversial. Critics question the government omnipotence in the management of regional water inequalities and the purely technical approach that governs arbitrations: the territorial, social, economic, and environmental dimensions, are often overshadowed.

On another level, the legal emptiness about water planning is a handicap to the development of planning instruments; this absence deprives the water sector of a comprehensive and prospective approach likely to promote sustainable water management. The conducting of studies, programs and strategic plans for the mobilization of water resources of the country can not constitute in itself a visible expression of the water policy and there is currently no single policy document outlining the main orientations of the water sector. Moreover, the National Water Council under the Water Code is rarely asked to give its opinion or to validate water programs. Finally, the consistency of the programs and their objectives are not subject to any public consultation, whether at national or regional level.

3.5.2 A Water Planning Model: Master Plans for Large Hydraulic Regions

Prepared by the implementation of the ten-year development prospects 1962–1971, water long-term planning is the backbone on which the water sector policy is articulated. The water infrastructure, designed in the Regional Master Plans (see Fig. 3.3), is based on the construction of large storage facilities, exploitation of groundwater, achieving great adductors for interconnecting reservoirs, and transporting water from surpluses basins to water deficit regions or to areas where opportunities for agricultural use are considered more favorable.

Fig. 3.3 The three major water regions covered by the regional water use master plans: North, Central, and South

The Northern Water Use Master Plan (PDEN)

The water available potential of PDEN is estimated at 2.5 km^3/year (55% of the country's total resources), including 1 km^3/year located at the Medjerda basin, the rest on other Northern basins. All sites suitable for the construction of large and medium dams, to enable mobilization of more than 85% of the potential of Northern surface resources, have been identified and reservoirs largely constructed, including the central structure corresponding to Sidi Salem dam near Beja: the keystone of national hydraulics system. The PDEN is based on the interconnection of large dams, which offers possibility of transferring water, reducing spills, storing maximum volumes, and improving water quality from mixing possibilities between waters of different salinities.

The Central Water Use Master Plan (PDEC)

The agricultural potential of the Central Region relies particularly on rain-fed arboriculture, and the rise in the use of irrigation due to blue water availability. The Western part of Central Tunisia is also the traditional source of drinking water supply for the coastal Sahel. The conception of the PDEC in 1976 implied that: (i) the main hydraulic facilities had to respect the internal logic of the hydrological systems in which they operate; for example, the Sidi Saad dam on Oued Zeroud should amplify the natural hydrological cycle, strengthening the Kairouan aquifer recharge; (ii) the water transfers to the Sahel do not antagonize the irrigation development in the source regions, so as not to accelerate population emigration. Consequently, water transfers to the Sahel should continue until year 2000 but beyond, it would be necessary to consider other water resources (desalination for drinking water, local resources for irrigation).

Overall, the PDEC available surface water resources were estimated at 320 million m^3/year. Groundwater was not yet known precisely, so estimated within a range of uncertainty from 320 to 450 million m^3/year, the withdrawals totaling then 220 million m^3/year. With intensive observations, groundwater is now better known and its resources have been reassessed at almost 700 million m^3/year, of which 650 million m^3/year are already withdrawn. This estimates change allows determining the scope and limitations of planning at the scale of a large hydraulic region: with data available at that time, the authors of the PDEC showed some lucidity in a prospective vision limited in scope to 20 years. Beyond this time interval, data and knowledge uncertainty prevails and planning needs to be updated.

The Southern Water Use Master Plan (PDES)

The PDES concerns the South pre-desert region, characterized by omnipresence of deep groundwater. The Northern Sahara aquifers allow an exploitable resource estimated at over 800 million m^3/year. With a contribution of surface water estimated at 200 million m^3/year, the total available water potential is at about 1 km^3/year. Management of this potential is however complex, due to: (i) the weakly renewable character of most available groundwater, (ii) its specific physical and chemical characteristics (temperature and high salinity for some), (iii) the need to implement very deep boreholes (sometimes up to 2500 m), often very expensive and sometimes distant from consumption centers. PDES has established a program consisting to: (i) satisfy domestic and industrial water needs, with a support of brackish water desalination to improve drinking water quality; (ii) safeguard 20,000 ha traditionally equipped for irrigation (more than one hundred oases) and develop 8000 ha of new oases. However, the achievement of these programs is today widely exceeded by private unauthorized extensions which result in massive overexploitation of a number of aquifers.

3.5.3 The Great Water Transfers of Tunisia

The context and conditions of water transfer

Tunisia always had to make efforts to overcome a very uneven spatial distribution of its water resources in terms of quantity and quality. The water allocation is often aggravated by a high interannual rainfall irregularity, and the demographic imbalance that has always favored the settlement of populations and economic activities on the east coast. To correct spatial inequalities and reduce resources inadequacies, large transfers were made possible by a water infrastructure articulated on large storage facilities, groundwater exploitation and achieving great adductors for reservoirs interconnection.

In addition to providing water to structurally deficit areas, water transfer also covers other objectives: (i) improved resilience of drinking water supply system to the risks of local deficits through inputs diversification; (ii) improved water quality by mixing waters from different origins.

The Dams interconnection strategy

The realization of inter basins transfers or within the same basin required the establishment of a fairly complex interconnection strategy which aims, in addition to achieving the above mentioned objectives, to transfer and store the surplus water to deficit dams, and to develop artificial recharge of overexploited aquifers with excess water from dams during wet years. The expected results of this strategy are to increase the total transfer capacity of surface water mobilized by dams, from 400 million m^3/year at present, to more than 900 million m^3/year in the long term.

Tunisia has ten large regional transfer systems illustrated by the scheme Fig. 3.4.

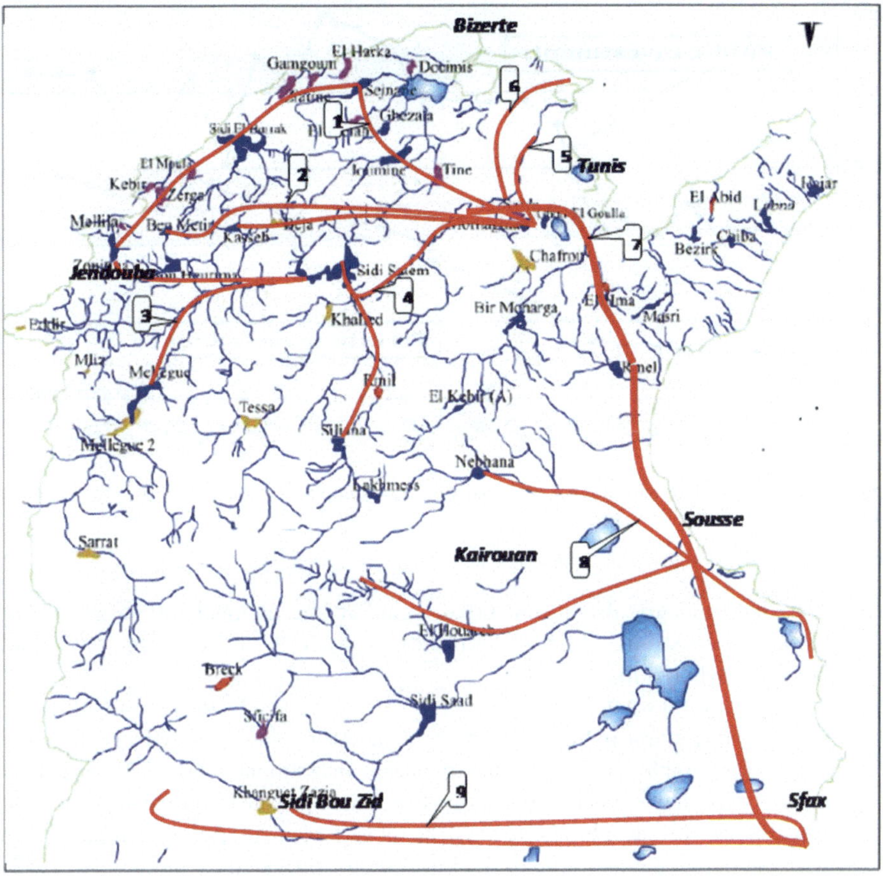

Fig. 3.4 Diagram of large water transfer systems

3.5.4 *Investments in the Water Sector*

Characteristics and evolution of water investments

Significant public investments were made, within the framework of the various plans for water infrastructure implementation: water resources facilities, irrigation schemes, drinking water supply, sanitation of urban areas, flood protection, etc.

Data from agricultural statistics and economic bulletins from the Ministry of Agriculture suggest the main following observations:

(i) Urban water has early occupied a prominent place in the Tunisian strategy by monopolizing 25–35% of water investments.
(ii) The share of agricultural water in water investment has always been relatively strong and stable, varying between 29 and 46%.

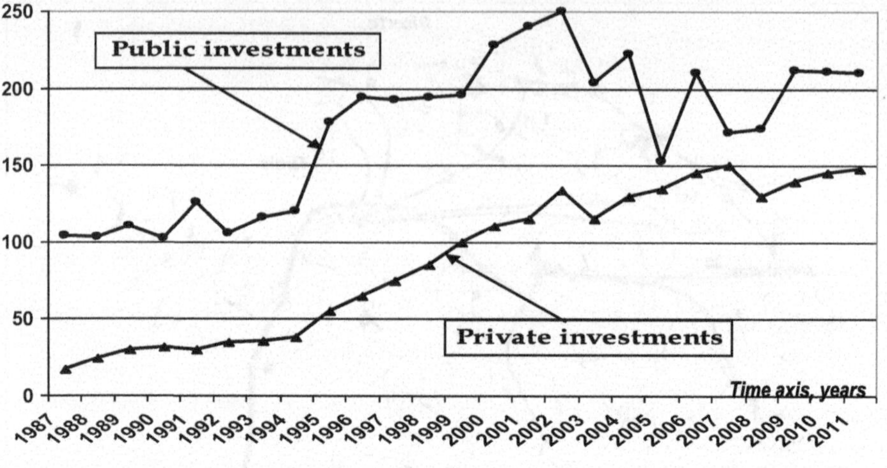

Fig. 3.5 Evolution of agricultural water investments, in million dinars (on 26/04/2017, the Tunisian Dinar was worth 0.38 € or 0.41 US $)

(iii) The share of agricultural and urban water in the national investment effort remains very strong (7–10%) until the construction of most large dams and large sewage treatment plants.

(iv) The agricultural water investments are strongly marked by the private initiative, which contribution increases from 14% in 1987 to 42% in 2010, with some years comparable to the amount of public investments (Fig. 3.5). For very positive it is, that marks the strong involvement of private initiative in productive investment, this development must be observed in its possible relationship with encouragements to the agricultural sector provided by the law N° 1982-67 (JORT 1982), concomitant increased groundwater exploitation observed since the mid-1980s, and the significant increase of private wells, an especially critical phenomenon since the early 2000s.

The Funding sources and mechanisms

The main partners of the Tunisian water investment policy are multilateral or bilateral funds contributors. The recourse to external financing is virtually systematic in the water sector, considered a capital intensive sector. The external participation is estimated in average at 60% of water public investment over the last three five-year plans. The demand for investment in the water sector remains strong to meet two key requirements: building new infrastructure and ensure the maintenance and rehabilitation of existing facilities. It is part of an international context marked by the debt crisis and a steady decline in official development assistance. A very small share of granted funding is intended to environmental protection of water resources, to general or strategic studies concerning the water sector, to training of operators and users awareness (irrigation water saving, support for user associations), and to R&D, all considered as fundamental aspects for a better water management.

Since the 70s, the bulk of financial cooperation has been devoted to water supply management and investment oriented specifically towards resource mobilization projects. The main partners in the implementation of this policy are the World Bank (WB), the African Development Bank (AfDB), the International Fund for Agricultural Development (IFAD), the Kreditanstalt Für Wiederaufbau (KFW), the French Agency of Development (AFD), the Japan Bank for International Cooperation (JIBIC), the Arab Economic and Social Development Fund (FADES).

The policy review and implementation of water sector reforms began in 1986 as part of a comprehensive structural adjustment program that gave priority to improve economic efficiency of the different sectors. This marks a new direction for contributors, which now favors the accompaniment of investment lending through structural reforms and measures related to water demand management and resource conservation: water pricing and cost recovery, associative management, groundwater management, fight against water pollution, capacity building, institutional and legislative framework. The PISEAU Project (Water Sector Investment Project, 2001–2014) is the expression of this new form of cooperation with the establishment by various contributors joint investment fund to similar or complementary objectives.

3.6 Institutional Organization and Intersectoral Coordination

The institutional framework based on public water management has allowed the implementation of water resources mobilization strategies. The centralized organization made it possible and has accompanied the policy related to water supply development in a regulatory framework conducive to a modern resource management. But this policy seems to be no longer fully compatible with a situation of total mobilization of water resources and its corollary, the context of scarcity and stress that characterizes today's state of water resources in Tunisia.

3.6.1 The Water Code and the Main Reforms

The legislative framework, administrative structures and available intervention mechanisms for water management are the means the Government uses to regulate the water sector. The latter is currently governed by the Water Code-Law No. 75-16-and by a number of texts relating to environment, sanitation and agriculture. Enacted in the mid-seventies JORT (1975), the Water Code was completely in keeping with the spirit of this time by introducing some principles, considered modern for the period, of rational water management and environmental protection. The Water Code is based on three main principles: centralization of the water administration, priority to drinking water and maximum resource valuation.

According to the Water Code, all water resources are considered as public good and as such are subject to State control. To release more the Government water action, in particular to promote large-scale irrigation schemes, the Water Code introduced a fundamental principle: realize the maximum value from each cubic meter of water across the country. In its economic meaning, the principle of maximum valuation leads to deliver better profit back of the water resource in various economic sectors paving the way for the possibility of water transfers that are well justified by recovery targets.

The Water Code provides for several measures to ensure the resource conservation and promote its efficient use, avoiding waste and degradation. When groundwater potential is deteriorating, the Water Code provides for production restrictions by creating Prohibition and Safeguard Perimeters or for uses regulation through Water Planning and Use Perimeters.

The water sector legislative and regulatory framework had been subject to readjustments. In particular, the Water Code amendment in 2001-Law No. 2001-216-led to a number of reforms by: (i) ensuring greater consideration of environmental requirements and promoting water savings, (ii) broadening the concept of water resources development to non conventional resources, (iii) integration of certain basic principles in the legal characterization of water resources: the adjective "national heritage" was attributed to water, integrating in a clear way the concept of sustainable development for water resources management. Indeed, the concept of national heritage refers to a certain value of the resource with all that this implies in terms of protection duty.

3.6.2 *The Institutional Framework and Intersectoral Coordination*

The Minister of Agriculture is the sole administrator of the Public Water Domain (DPH), assisted in this regard by two advisory bodies: (i) the National Water Council, an advisory council which provides advice on national plans and large water projects; (ii) the Commission of DPH, produces technical advice on any question relating to the Public Water Domain administration.

The administrative organization of water at central and regional levels essentially revolves around specialized agencies under the Ministry of Agriculture. At the central level, some branches of the Department assume technical tasks related to planning, monitoring-evaluation and control of water resources, and direct management of large hydraulic structures. At the regional level, the Regional Commissions for Agricultural Development manage water resources: they implement and coordinate the government water and agricultural policy within the governorates boundaries. The device of the Ministry of Agriculture also includes autonomous public institutions: the National Water Exploitation and Distribution Company (SONEDE) is responsible for drinking water supply in urban areas and some rural communities;

the Northern Water Canal and Adductions Exploitation Company (SECADENORD) is responsible of large transfers of Northern water systems.

To encourage water management decentralization, Agricultural Development Associations (GDA) manage water resource for irrigation and water drinking supply in rural areas at the local level. In 2010, there were 2450 GDAs including 1250 for drinking water management in rural areas, 1050 for irrigation and 150 GDAs for both. Irrigation GDAs manage about 85% of public irrigated areas.

Although the Ministry of Agriculture is vested with responsibility for water resources management, other ministerial departments are involved, within their prerogatives, in the Public Water Domain: The Ministry of Health is responsible for the water health aspects, the Ministry of Equipment is involved in urban protection against floods, the Ministry of the Interior is involved in the management of flood events and coordinating the various interventions. The Ministry of the Environment has, among its missions, water resource conservation, particularly the fight against pollution of aquatic environments and control and monitoring of water pollution.

3.7 Priority Issues and Risk Management

Describing the status of water uses and management methods in the different sectors allows identifying difficulties and limitations of implemented strategies with the aim of putting into perspective the water-related guidance.

3.7.1 Drinking Water and Sanitation

Drinking Water Services

Founded in 1968, SONEDE had the task of mobilizing, producing, treating and distributing drinkable water. Access to public water services was so paltry (Chabbi 2006): in 1966, ten years after Independence, only 15% of the population had safe water at home and the rate of public sanitation networks connection did not exceed 5%. In 2010, drinking water supply rate is estimated at 100% in urban areas and almost 94% in rural areas. A portion of the rural population in the agglomerated areas (49%) are served by SONEDE, but a no less important part (45% of the rural population, located in dispersed areas) are supported by the Directorate General of Rural Engineering and Water Exploitation (DGGREE), which provides infrastructure and equipment operated by drinking water GDAs. Translated not in terms of service, but in terms of individual connection rate, this indicator rises to 99% in urban and 45% for rural populations.

Since the creation of SONEDE, the performance displayed by the sector is edifying, including during the two last decades (Fig. 3.6): drinking water production increased from 90 million m^3 in 1968 to 580 million m^3 in 2012; and between

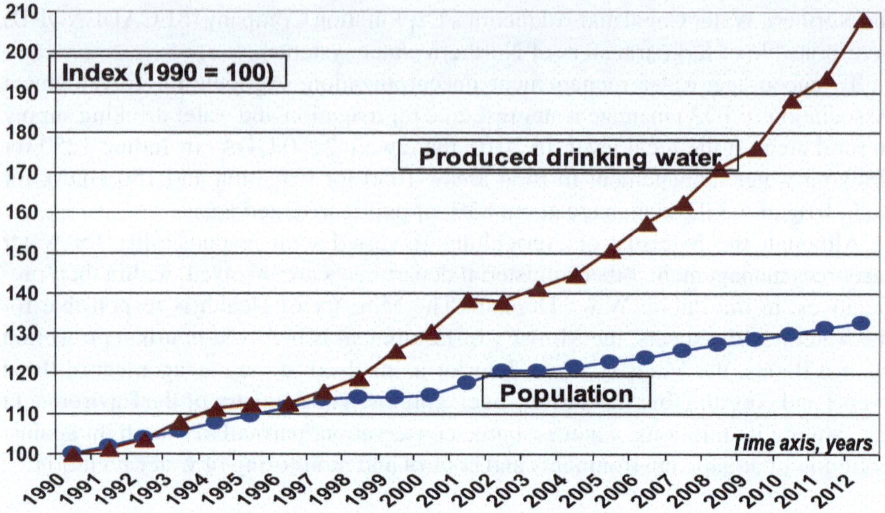

Fig. 3.6 Evolution of Tunisian population and produced drinking water volumes (data from SONEDE)

1975 and 2010, the drinking water specific consumption increased from 35 to 90 L/day/capita on national average. However, as shown in Fig. 3.7, these national averages do not reflect regional differences, and five regions (governorates) are still below the 55 L/day/capita, those of Jendouba, Siliana, Kairouan, Sidi Bouzid, Kasserine, regions where the presence of rural population is precisely the highest.

So, notwithstanding all the efforts made by the country in the field of water mobilization, the priority given to drinking water since 1975 by the Water Code, the relative progress in drinking water supply to rural areas, and all the displayed performance, significant progress remains to be made: indeed, a million and half rural people, about 15% of the Tunisian population, are still supplied with drinking water by collective systems and can not have water directly at home. On the other hand, 220,000 rural people had not yet (statistics in 2012) access to safe drinking water.

The sanitation services

Created in 1974, ONAS is the nationwide operator responsible for wastewater collection and treatment. Of a total population of 6.3 million living in the municipalities supported by ONAS, 5.6 million people are now connected to the public sewerage networks. In these cities, all urban networks can collect a total volume of 246 million m^3/year (ONAS 2010) of waste water almost entirely (240 million m^3/year) treated in 110 treatment plants of different sizes spread throughout the country. In the sanitation sector as in the drinking water one, these positive country-wide average figures conceal significant differences between the regions characterized by high population

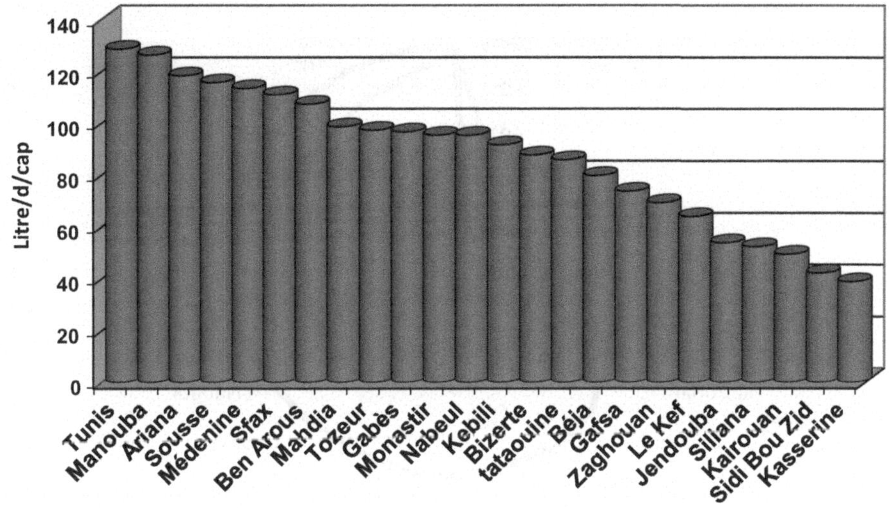

Fig. 3.7 Specific drinking water consumption in 2011, regional values classified by Governorate, L/day/capita (data from SONEDE and DGGREE)

density and economic activity concentration and the rural and remote areas located far from the main centres of economic activity (Fig. 3.8).

The problem of connection to drinking water and sanitation in rural areas arises today as one of the major challenges of the water sector. While significant progress has been made in the field of drinking water, rural sanitation problem remains complete. Institutionally, rural sanitation does not fall directly under the prerogative of ONAS even if ONAS is involved on an ad hoc basis and on behalf of the government in some rural agglomerations, in particular through pilot programs. If real progress has been made on the technical level, the institutional aspects of rural sanitation have not been included in the pilot programs and from this point of view, this remains an open ended question.

3.7.2 Water and Agriculture

The total water withdrawals are 3.0 km^3 in 2010, 2.3 km^3 of which are allocated to irrigation, or 77% (17% devoted to collectivities, 3% to industry and 3% to aquatic ecosystems). 77% of irrigation demand comes from groundwater, 22% from surface water, and 1% from treated wastewater reuse.

Given the water resources allocated to agriculture, the irrigated potential area is estimated at 560,000 ha, including 410,000 ha in total or partial control and 150,000 ha in supplemental irrigation and flood spreading.

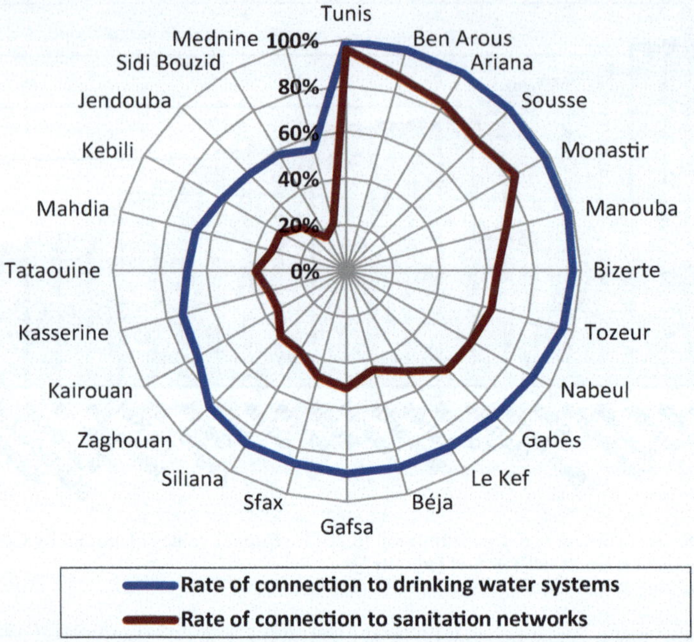

Fig. 3.8 Connection rate to urban drinking water and sanitation networks by governorate in 2010 (data from INS 2011)

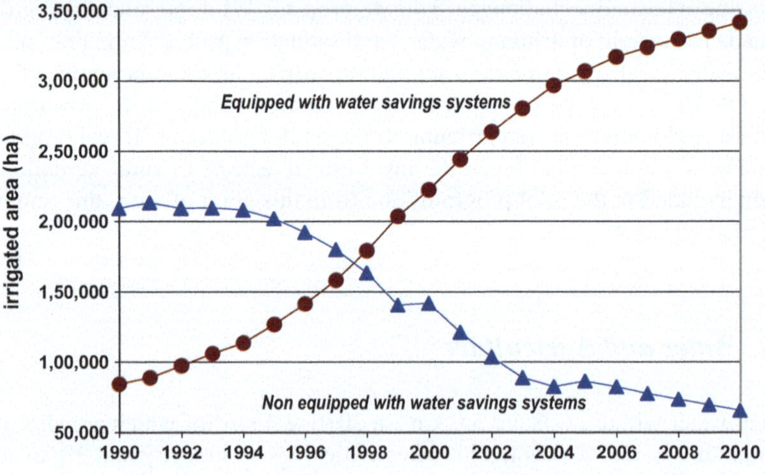

Fig. 3.9 Evolution of irrigated areas equipped with water saving systems (DGGREE 2012)

The efficiency of collective irrigation networks is estimated at 85%. An extensive water-saving irrigation program started in 1995, and the rate of equipment with water saving systems reached 85% of irrigated land in areas under total control in 2014. This

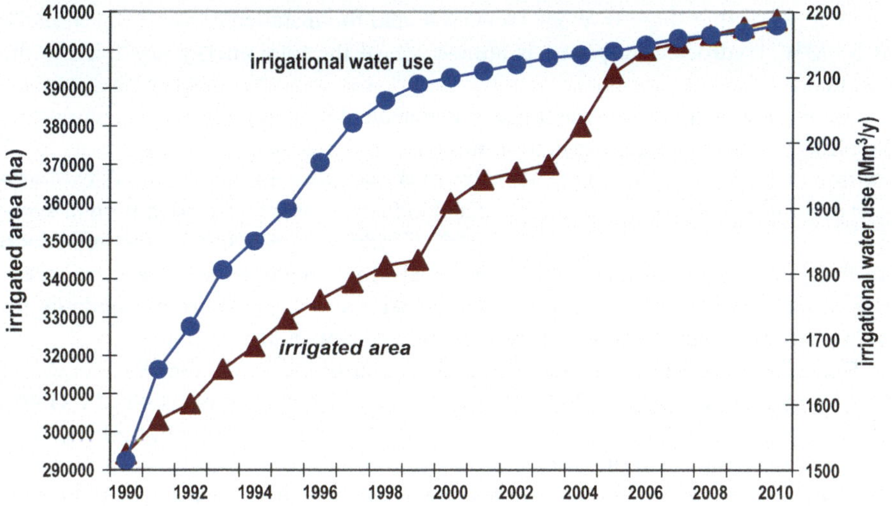

Fig. 3.10 Irrigated areas and irrigational water use from 1990 to 2010

rate was only 37% at the start of the program in 1995 (Fig. 3.9). Currently, nearly all the area under irrigated agriculture in Tunisia is equipped with pressure pipe networks operating on an on-demand basis for the largest part and with aspersion, or localized parcel irrigation systems on 75% of Public Irrigation Schemes. This strategy had allowed, in fifteen years, for a certain stabilization of irrigation demand despite the extension of irrigated areas (see Fig. 3.10).

Although irrigated agriculture is practiced on only 8% of the country's agricultural land, it accounts for up to 37% of the total value of the agricultural production. This potential for value and job creation in rural areas is a major opportunity that should be built upon to contribute to the economic and social development of the country's regions, especially the most deprived of them (regions considered as underprivileged). Developing irrigated agriculture would therefore perfectly align with the country's key objectives, i.e. revitalize economic growth, increase productive investment, reduce unemployment, and alleviate poverty, particularly in rural areas.

Furthermore, excessive use of groundwater in some regions severely threatens the sustainability of the current development of irrigated lands.

3.7.3 Energy, Industry and Tourism Water Uses

Tunisia has no major rivers or sites suitable for hydropower continuous production. Since the 50s, several small projects were carried out in connection with the construction of reservoir dams in the North. Currently, the national hydroelectric potential is estimated at 1000 GWh. The technically exploitable potential is 250 GWh, the economically feasible share is about 160 GWh, and the installed capacity is estimated at 66 MW. Hydroelectric power is almost 1% of the total energy produced in the country.

As for data on industrial water use, it is managed by two public organizations: (i) SONEDE controls water distribution to connected industries which amounts to 35 million m^3/year in 2010; (ii) the Ministry of Agriculture, through aquifers exploitation periodic records, controls unconnected industries, which use 60 million m^3/year the same year. Since water volumes used by industry are generally moderate, and abstraction on groundwater remains relatively low, manufacturers have no incentive to introduce water-saving processes. There are not yet Tunisian standards for industrial uses, and recycling is practiced only by few industrial units.

Concerning Tourism, certainly individual consumption of tourist (Average of 550 L/day/bed occupied in 2006; 900 L/day/bed in 5-star hotels (Lahache Gafrej 2007) is very high. On average, it is six times the daily consumption of Tunisian domestic user. However, the total consumption of the entire tourism sector reaches 25 million m^3/year which represent less than 1% of total water uses, including the agricultural sector, and 5% of drinking water uses.

3.7.4 Non Conventional Water Resources

To improve its water balance, Tunisia has sought since 1970s to increase its water availability by non conventional means: reuse of treated wastewater and desalination of brackish water and seawater.

The reuse of treated wastewater

Due to the nature of treatment processes applied to wastewater (secondary type), the treated wastewater reuse (TWR) is authorized in Tunisia only for agricultural purposes. For irrigation, this reuse obeys to specific rules on users' health protection and on agricultural products. TWR is only authorized for food products which may not be eaten raw, for industrial crops, forestry and horticulture.

However, TWR does not follow the sanitation sector performances: many technical, socio-economic and cultural reasons still hinder the effective reuse of waste water, which is barely 10% of treated wastewater potential. The technical reasons are related to the quality of water produced and the restrictions imposed by the environmental and health regulations. Other technical difficulties deal with the conditions of transfer and control used to deliver treated water to areas of high consumption and adapt the pace of its production to that of its use. Finally, and despite incentive pricing, TWR encounters a strong users' reluctance.

Despite these difficulties, TWR has helped maintain or develop agricultural activity in areas with locally structural deficiency of conventional resources. Successfully applied in golf courses irrigation, TWR has reduced pressure on an already stressed resource in the touristic coastal areas. Reuse also limits releases in vulnerable natural environments helping to reduce impacts and achieve at the same time important savings in water treatment, such as: (i) in areas where tourism is important, where more and more use is being made of costly sea outfalls to disperse residual pollution of treated water away from the beaches, (ii) in Medjerda basin where complementary urban sewage treatments are used for nutrient removal in order to reduce the risk of eutrophication of rivers, reservoirs and lakes.

Desalination

Desalination of brackish water (deep groundwater, salinity between 3.5 and 6 g/L) is practiced by SONEDE in Kerkennah island (3300 m^3/day), Gabes (22,500 m^3/day), Djerba and Zarzis (24,000 m^3/day each), at a cost of about 0.800 DT/m^3 (SONEDE, 2006).The volume of produced desalinated water reached 18 million m^3 in 2007, about 4% of the total drinking water production. Water desalination has become a strategic option for SONEDE to improve drinking water quality for cities in the southern region of the country. Some tourism and private industrial units already resort to desalination of brackish water or sea water for their own account.

Moreover, the principle of equity in the access to drinking water have led to distribute and use at the same cost and practically in the same technical conditions, water irrespective of its origin including water produced by desalination. Surely that is an expression of solidarity at the national level for strictly domestic use, but that does not necessarily lead to the best conditions of water resources conservation and economic valorization.

3.7.5 The Hydrological Risk Management

Risks may originate from natural hazards (droughts, floods) or result from human action, often in relation to management and use of the resource: water and soil salinization, reservoirs silting up, risk for aquatic ecosystems, water pollution.

Flood management

In the recent past, the 1969 floods were the most devastating on the center and south of the country, those of 1973 struck the North causing considerable damage especially in the Medjerda lower valley. The city of Sfax in the South was in 1982 severely affected by floods with economic consequences and damage in terms of human losses. The city of Tunis has also experienced severe flooding in 2003. One can multiply the examples. Actually, flood management is not given special status: the general status of natural hazard management applies by default to flooding. Under Act 91-39, floods are considered as natural disasters when their severity and the effects they generate are beyond the ordinary means to deal with them. The prevention of natural disasters and response measures at the national level are organized in the framework of a national plan and supervised by the "Permanent National Commission for Fight against disasters, prevention and relief organization", under the Ministry of Interior authority. At regional level, regional commissions placed under the authority of the governors perform such missions as part of regional plans.

Droughts

The great droughts of the twentieth century (1937–1938, 1947–1948) are still in memories: they were devastating, causing severe famines. The storage dams now help ensuring interannual regulation of surface water resources and mitigating the effects of drought. But prolonged droughts are a reminder that Tunisia is an arid country whose resources are structurally insufficient. The socio-economic damages of droughts remain extremely high, especially at the level of the agricultural production and the trade balance.

The Water Code does not address issues related to drought. However, in the event of drought, the Administration must declare a situation of shortage, defines and delimits the affected areas and enacts local and cyclical regulations, including water rationing measures, in order to ensure priority in providing drinking water for populations, feeding livestocks, and where possible safeguarding of existing plantations.

Irrigation and soil salinization

More than half of mobilized water allocated for irrigation is considered to varying degrees as "poor" to "bad" for irrigation. While 72% of mobilized surface waters have a salinity below 1.5 g/L, only 20% of groundwater fall into this category. The brackish water management rules at the agricultural plot level, that would reduce the risks of salinization, are known and disseminated. Their conditions of application are met in most public irrigated areas, but all the irrigators do not necessarily observe good irrigation practice instructions or underutilize water available volumes to ensure adequate leaching salts. The drainage system does not always receive the necessary maintenance, and in some cases where natural drainage is deficient and in the absence of artificial drainage network, the eventually uplift of groundwater level leads to soil salinization. The Soil Department of the Ministry of Agriculture develops a monitoring network on both: (i) soil salinization in public irrigated areas at risk, (ii) salinity and water levels of groundwater in areas that may be affected by water logging due to irrigation excess.

Siltation of dam reservoirs

In Tunisia, the greatest threat to surface water mobilization works is premature silting of dams and hill lakes reservoirs. It has been estimated by Ben Mammou and Louati (2007) that volumes lost by sedimentation should be 500 million m^3 in 2010 and reach over 1100 million m^3 in 2030, nearly 40% of the storage capacity of existing dams. A vast program of water and soil conservation is already committed to address in the long term nearly 3 million hectares affected by erosion and to reduce risk of reservoirs silting. But in this field, as in many others, financial investments often granted under emergency conditions, are not followed by objective evaluations of programs effectiveness.

Environmental risks and water ecosystems protection

The realization of a multitude of projects has led to a sharp water cycle artificialization. In these circumstances, the environmental protection requires to be systematically considered in the resource allocation programs: releases to ensure water flows for lacustrine environments, wetlands and groundwater recharge. This perception of environmental demand has been gradual, and became an essential part of water management methods (Saied and Elloumi 2007), the most demonstrative example being that of the Lake Ichkeul in the North: the master plan for water in Northern Tunisia has been updated to integrate the lake Ichkeul as a "water consumer", and number of large dams now include the function of ensuring the environmental demand.

Pollution risks and water resources quality monitoring

The Water Code, enacted in 1975, provided systematic inventory of the surface water "degree of pollution" without providing implementing legislation specifying modalities of such an inventory. Later, water pollution sources inventories, mainly industrial and urban sources, were realized by the Ministry of Agriculture in 1994, and by the Ministry of the Environment in 2004, which identified about one thousand water pollution sources. On the risks induced by human activities, the level reached by a number of indicators remains very worrying for the future of water resources.

To conclude on hydrologic risks in Tunisia, and despite considerable investments on the last fifty years, both in financial and cognitive terms, the real impacts of massive water mobilization and management strategies are not yet precisely identified: decrease of groundwater reserves, actual impacts of water saving measures in irrigation, reuse possibilities of treated wastewater, irrigation impacts on soil and groundwater salinization, impacts of water and soil conservation works on the protection of dam reservoirs or groundwater recharge.

3.8 Limits of Water Policies and Challenges Ahead

3.8.1 An Exploitation of Water Often at the Limits of Resources Potential

Studies on the water sector conducted over the past two decades have identified the main problems and specified the appropriate management modes of water scarcity. But their conclusions had not affected the fundamental choices of the national water policy that are to further develop water resources exploitation in order notably to increase agricultural production and ensure the food challenge. Rationalization of water use in agriculture was a key part of the water management strategy and significant financial incentives have led to a significant extension of water saving equipments.

The National Program for Water Economy allowed to pass up to threefold, between 1995 and 2010, the areas irrigated by modern efficient equipments. Certainly, investments in this context helped to increase the overall efficiency of irrigation, but overall agricultural water consumption does not seem diminishing; the amounts of water saved are eventually reallocated to increase irrigated areas, particularly in deprived zones. The irrigation schemes programs have pushed the logic of irrigation development to levels that appear incompatible with the already high levels of water resources exploitation in number of water table aquifers. In this context, the government planned to continue efforts to achieve a mobilization rate of 95% in 2016 (large dams programs, their interconnection and the transfer of surpluses, new medium-sized dams and hill lakes, boreholes) and to strengthen the development of non conventional resources. This closer management of the resources, almost completely mobilized, leaves little room for environmental considerations.

3.8.2 Strengths and Weaknesses of the Mobilization Works Investment Policy

The Investments in water infrastructure mobilization or in the development of large irrigated areas account for decades for a significant proportion of funds allocated to the Ministry of Agriculture. Moreover, the funding policy of water mobilization and development hides a number of disorders, which may jeopardize the sustainability of hard-won gains: (i) reduction of reservoirs capacity and groundwater overexploitation; (ii) intensive use of weakly renewable groundwater resources in the South.

In the North, a mass of considerable investment has been earmarked for the development of large public irrigation schemes in areas where irrigation does not always seem to be of higher profitability. Certainly the irrigation option for Tunisia represents a fundamental choice, but the level of agricultural development in these areas is still modest in most cases. The average rate of crop intensification (ratio of the total area actually irrigated during the campaign to the total irrigable area) is variable depending on rainfall of the year, but remains relatively stable over the past decade, around the order of 90%, which may vary between 40 and 120% depending on the scope. The potential intensification rate of irrigation is estimated at 120–130%, allowing in principle an area available for irrigation of more than 500,000 ha for the whole country with the same water allowance. This paradox relating to the under-utilization of irrigation schemes in an arid country causes underutilization of about the third of surface water resources that in reality strengthen large volumes of water in reservoirs in average hydrological years, to use it the years of shortage: This phenomenon, which has prevailed over the last decades, is fading due to increased pressure on water resources in recent years.

Water mobilization is in a near-saturation phase and it is important that the intensive investment policy in hydraulic infrastructure be reassessed and readjusted with a view of valuating existing facilities. On the other hand, funding, becoming increasingly rare, should be released for other priority needs, particularly the conservation of water and soil in the fight against silting, maintenance and rehabilitation of equipments, agricultural extension, training and research, economic infrastructures.

References

Aubert F (1892) Carte géologique de la Tunisie au 1/800000ème

Bargaoui Z (1993) Les débuts de l'hydraulique moderne en Tunisie; Al Madar N° 1, L'homme et la maîtrise de l'eau en Tunisie; Ed. Cité des Sciences de Tunis

Ben Mammou A, Louati MH (2007) Évolution temporelle de l'envasement des retenues de barrages de Tunisie » Revue des sciences de l'eau. J Water Sci 20(2):201–210

Burollet F (1995) Exploration de la Tunisie avant la 1ère guerre; Comité français d'histoire de la Géologie 3(tome 9):111–122

Chabbi M (2006) L'accès aux services essentiels en milieu urbain; le cas de la Tunisie; Institut de Développement Durable et des Relations Internationales, IDDRI

Chevalier L (1950) Le plan de modernisation et d'équipement pour l'Afrique du Nord. In: Population, (french edition), vol 5, no. 1, pp 166–170

DGGREE (2012) Rapports annuels du système de suivi de l'exploitation des périmètres publics irrigués

DGRE (1990) Stratégie pour le développement des ressources en eau de la Tunisie au cours de la décennie 1991-2000; Septembre 1990

DGRE (1999) Etude du secteur de l'eau. Groupement Bechtel./Scet -Tunisie

El Amami S (1984) Aménagements hydrauliques traditionnels de Tunisie. CRGR, Tunis

Gauckler, P. (1897). Enquête sur les installations hydrauliques romaines en Tunisie, Imp. La rapide, Tunis

Gosselin M (1952) L'inventaire des ressources hydrauliques de la Tunisie. Imprimerie Nationale, Paris

Gosselin M (1969) Aperçu de l'évolution de l'hydrogéologie pendant les quarante dernières années, Congrès National d'Hydrogéologie. Journées Henri Schoeller, Bordeaux

Hamza M (1992) Un siècle de reconnaissance et d'exploitation des aquifères par sondages en Tunisie. DGRE, Tunis

INS (2011) Rapport annuel sur les indicateurs d'infrastructure, année 2011. Institut National des Statistiques, Tunis

JORT (1975) Law No. 16-75, of March 31, 1975, promulgating the Water Code. Journal Officiel de la République Tunisienne

JORT (1982) Law N° 82-67 of 6/8/82 encouraging investment in agriculture and fisheries. Journal Officiel de la République Tunisienne

Lahache Gafrej R (2007) Comment réduire la consommation d'eau dans le secteur touristique en Tunisie, approche et stratégie; Water demand management in the Mediterranean, progress and policies. Blue Plan, Zaragoza

Morin P, Memmi L (1972) Bibliographie analytique des Sciences de la Terre - Tunisie et régions limitrophes. C.N.R.S-CRZA, Paris

ONAS. (2010). Rapport d'activité 2010

Saied M, Elloumi MJ (2007) Prise en compte des besoins écologiques de l'Ichkeul dans la gestion de l'eau, Gestion de la demande en eau en Méditerranée, Zaragoza

Solignac M (1936) Travaux hydrauliques hafsides de Tunis, Revue Africaine No. 368-369, 518-580

State Secretariat for Planning and Finance (1961) The ten-year development perspectives 1962–1971. Tunis

Thomas P, Pervinquière L (1907, 1908, 1913) Essai d'une description géologique de la Tunisie » en 3 tomes

Tixeront J, Berkaloff E (1954) Recherches sur le bilan d'eau en Tunisie. UGGI-AIHS, assemblée générale de Rome
WWAP (2009) The United Nations world water development report 3, case studies volume: facing the challenges. World Water Assessment Programme
Zamiti K (1985) La division du travail étatique: sociologie d'un barrage, Annuaire de l'Afrique du Nord, 22, 377–388, K edn. CNRS, Paris

Chapter 4
The National Water Balance

Contents

4.1 Introduction

This chapter proposes to establish a national water supply and demand balance, both understood in their broadest sense: including the offer of natural rainfall throughout the country, comprehensive demand including needs of society, economy and the environment. Water quantities required for food production are not covered in this

M. Besbes et al., *National Water Security*,
https://doi.org/10.1007/978-3-319-75499-4_4

chapter, which is essentially devoted to the blue water balance. The national balance of green water and virtual water will be discussed in detail in Chaps. 7 and 8 but a consistent introduction is dedicated here to the soil water resource estimation because of its great importance in the national balance.

The first part (Sect. 4.2) of the chapter describes the general hydrological context of Tunisia. The second part (Sect. 4.2), from an historical perspective and knowledge of the Tunisian water balance, presents the general state of water resources: first the generic rainfall resource, then the state of surface water and groundwater resources, with a focus on weakly renewable resources and an update of the exploitable resources taking into account the water salinity criterion. The reserve provided by water in the agricultural soil constitutes the last great natural accumulation of water resources. This part ends with a brief review of climate change impacts on the resource. The third part (Sect. 4.4) focuses on the exploitation of water resources and presents updated information on water facilities and developments such as large dams, hill lakes and hill dams, water-table and deep aquifers, but also Evapotranspiration, called "green water", and touches on enhancing surface water by water and soil conservation.

The fourth (Sect. 4.5) reports on the different uses of water: for drinking, industry, agriculture, environment, and describes the changing resource-use balances that maintain regional imbalances with large inter-basin transfers, presented in the fifth part. A sixth part (Sect. 4.7) discusses groundwater overexploitation, which represents the most terrible threat to the future of water resources in the country, a threat likely to significantly and sustainably transform the national water balance. The seventh (Sect. 4.8) is devoted to the anthropogenic water cycle, where we attempt to provide a precise assessment of all water use: from withdrawal, through the various stages of allocation and actual use, to net consumption and returns to the receiving environment. The last part tries to establish the complete water balance of Tunisia, where we propose a first prognosis of the future of all rain-fed resources.

4.2 The Hydrological Setting

With an area of 164,420 km^2 and a population of 10.7 million inhabitants in 2011, Tunisia (Fig. 4.1) has rainfall resources estimated at 36 km^3/year. Total water resources (blue water) are estimated at 4.85 km^3/year, of which 2.7 km^3/year are the average runoff or surface resources. In 2010, groundwater withdrawals were estimated at 2 km^3, representing an average operating rate of 93% of exploitable groundwater resources estimated at 2.15 km^3/year, but this average hides large regional differences and many aquifer systems are under severe overexploitation.

The water resource potential of agricultural rainfed soils (green water), available for evaporation and consumption by plants, related to arable land (5 million ha), is estimated in average at 13 km^3/year. This potential increases to nearly 19 km^3/year if one incorporates rangelands (5 million ha).

The total blue water withdrawals are estimated at 3.0 km^3 in 2010, of which 0.54 km^3 are allocated to drinking water (55% surface water and 45% groundwater)

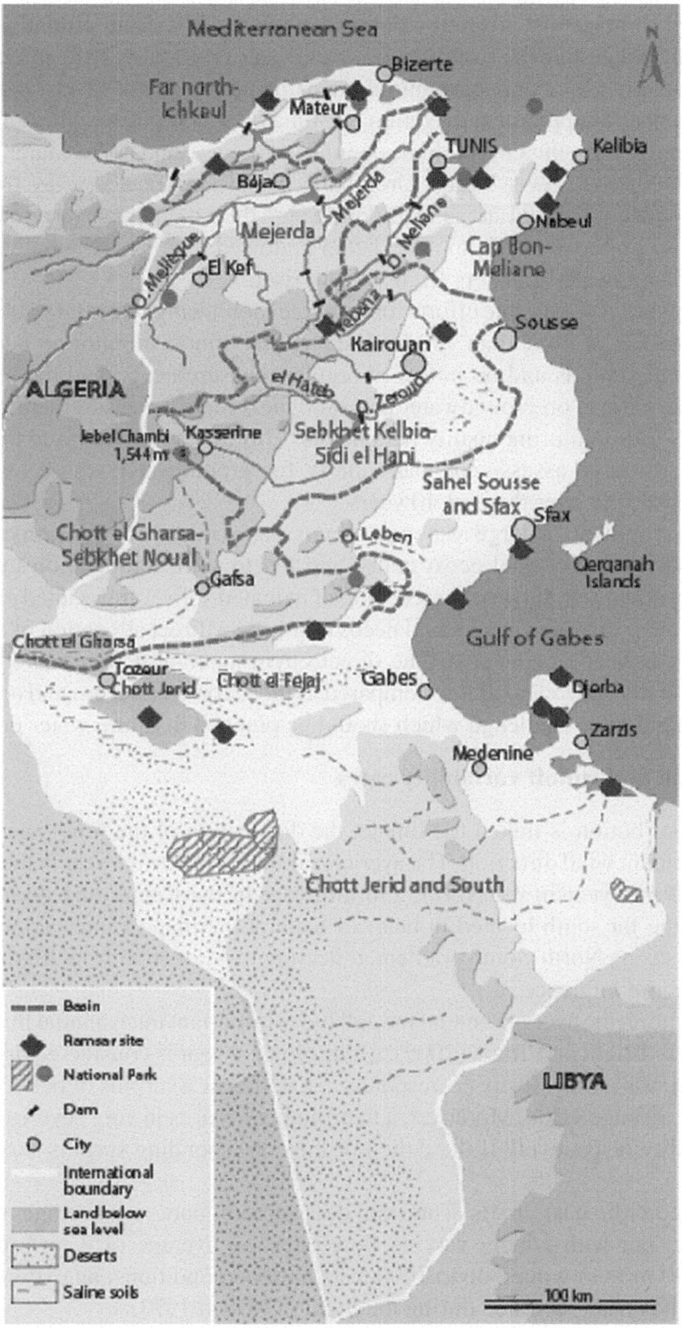

Fig. 4.1 General Map of Tunisia, with main basins, regions and cities. *Source* World Water Assessment Programme WWAP (2009)

and 2.3 km^3 to irrigation. Irrigation demand comes to 77% from groundwater, 22% from surface water and 1% from treated wastewater reuse. Given the affected water resources in agriculture and agricultural intensification possible levels, the potential for permanent and supplement irrigation is estimated at 560,000 ha. Most of the water demand comes from the coastal high populated regions, and some major irrigation schemes (Medjerda Lower Valley, Grombalia, Sahel) are located away from major resource deposits. Coastal areas use more water than their local resources, which leads them to import water from other better endowed regions: the whole country is marked by large water transfers from West to East.

This requires considerable efforts for resource monitoring: the rainfall observation network consists of more than 900 regular stations; runoff monitoring network has 60 permanent stations and 60 points of regular measurements; and the network of regular observations on groundwater has 3800 piezometric measurement points and 1200 points for monitoring quality parameters. This information is used to regularly update the resource assessment, particularly for groundwater whose exploitation increased by 330% over the past 40 years.

The development of large water programs has reduced inputs to hydrological systems, inland and coastal ecosystems, and led to some salinization of irrigated soils. Water quantities allowed for leaching of irrigated soils were granted early while the definition of other environmental needs requiring a direct allocation of mobilized resources was later to conceive. The direct environmental demand (wetlands and artificial recharge), while still low compared to those for urban areas and agriculture, is an emerging need challenge which should be planned in future water programs.

The rainfall and runoff variability rates

Rainfall distribution is linked to climate, the disposition of the reliefs and the NW winter dominant wind direction. The average rainfall is 1500 mm/year in the far north, around 500 mm/year in the North, 250 mm/year in the Centre and do not exceed 50 mm in the far south located in heart of Sahara: the map of inter-annual rainfall indicates a strong North-South gradient, reflecting the transition from Mediterranean area to Saharan influence.

Furthermore, the average annual rainfall hides important inter-annual fluctuations. According to Frigui and Touzi (2009), a hydrological year is considered dry or damp when the rainfall respectively corresponds to a deficit or a surplus of between 30 and 50% of the average value. Moreover, a hydrological year is in very severe drought or high humidity respectively if the deficit or the corresponding surplus exceeds 50% of average.

The Medjerda has, in its Tunisian downstream part, an average supply of 1 km^3/year, but with widely varying flow rates: an average of 3 m^3/s in the dry season to 90 m^3/s in winter, up to 1000 m^3/s in flood conditions; an extreme flow of 3500 m^3/s has been recorded during the flood of March 1973.

Basins of the extreme North (Oued Barbara, Melilla, Kebir Zouara) and Ichkeul (Joumine, Rhezala, Sejenane, Douimis) extend over a total area of 5000 km^2 and have a remarkable average intake of 0.96 km^3/year, due to a well-stocked rainfall.

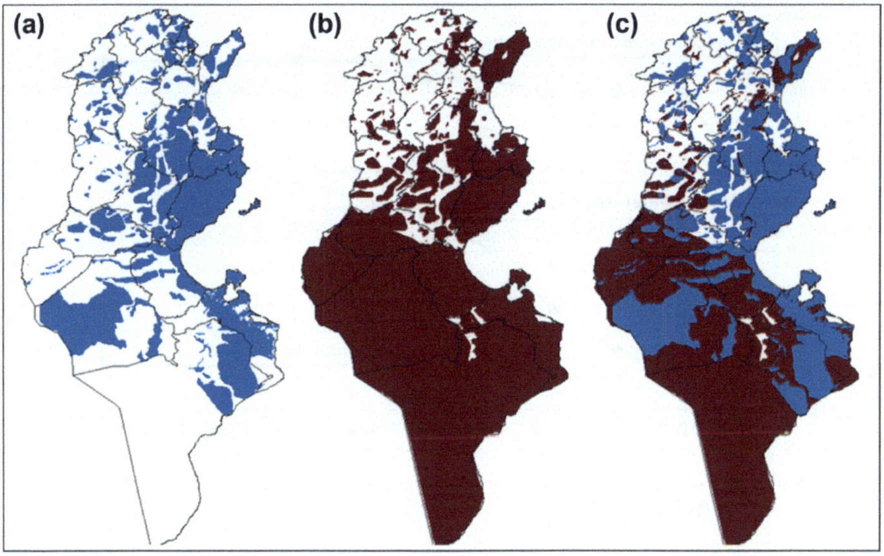

Fig. 4.2 Extension of aquifers: **a** shallow aquifers; **b** deep aquifers; **c** superposition of shallow and deep aquifers. (Author's elaboration according to data from DGRE)

Spatial distribution of aquifers and groundwater in tunisia

Tunisian hydrogeologists conventionally distinguish phreatic aquifers exploitable by large diameter wells, usually constructed of masonry, and deep aquifers exploitable by drillings on several hundreds meters.

Figure 4.2 shows the spatial distribution of Tunisian aquifers. The Water Resources Department (DGRE 2005, 2007a) counts 210 shallow and 195 deep aquifers. The extension of these aquifers represents 120,000 km^2 area: three quarters of Tunisia are aquifers.

4.3 The State of Water Resources

4.3.1 Knowledge and Evolution of the National Water Balance

After several decades of studies and research, and a continuous evolution of the water balance terms, the knowledge of water resources, mainly groundwater, corollary to the exploitation level and analysis of monitoring networks, appears imperfect. This is emphasized in Fig. 4.3 which presents the developments of successive estimations respectively of surface water resources, exploitable groundwater resources, and the groundwater withdrawals between 1954 and 2010.

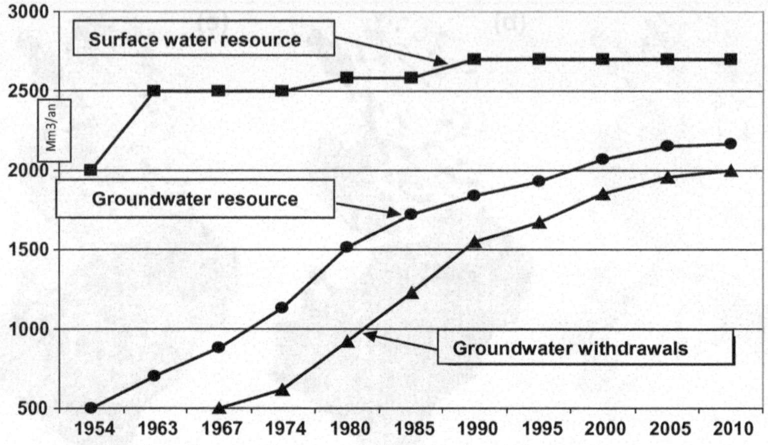

Fig. 4.3 Evolution from 1954 to 2010, of surface water resources, groundwater resources estimates and groundwater withdrawals

If for surface water where flows are visible and measurable, it has been early access to plausible orders of magnitude, where one observes a certain stabilization of the estimated values, it is not the same for groundwater, for which present value of resources represents more than four times initial estimates. This deals of course with valued resources whose growth reflects the knowledge progress of withdrawals. This development shows how much reduction in estimation uncertainty is strategic in Tunisia, where the "resource-needs" budget is so tight. This clearly reflects the need to invest more and more in knowledge of the elements that determine the groundwater resource.

Moreover, the major observed or expected hydrological changes (national hydrological cycle almost totally artificialized, climate change) require also continuous review on precipitation and surface runoff knowledge.

4.3.2 The Rainwater Resources

Rainwater resources are defined as the amount of rainfall measured in average hydrological year across the whole Tunisian territory. Calculated on a series[1] of nearly 100 years, covering all the regions from 1916 to 2010, the average area-weighted rainfall of Tunisia amounts to 220 mm/year, with a maximum of 387 mm and a minimum of 127 mm. This historical series shows no long-term trend evolution (Fig. 4.4). This corresponds, for an area of 164,420 km^2 including

[1]Representative synchronous observed serials for the 24 governorates; rainfall weighted averages according to annual values communicated by Mustapha Saadaoui (DGRE), personal communication to the author.

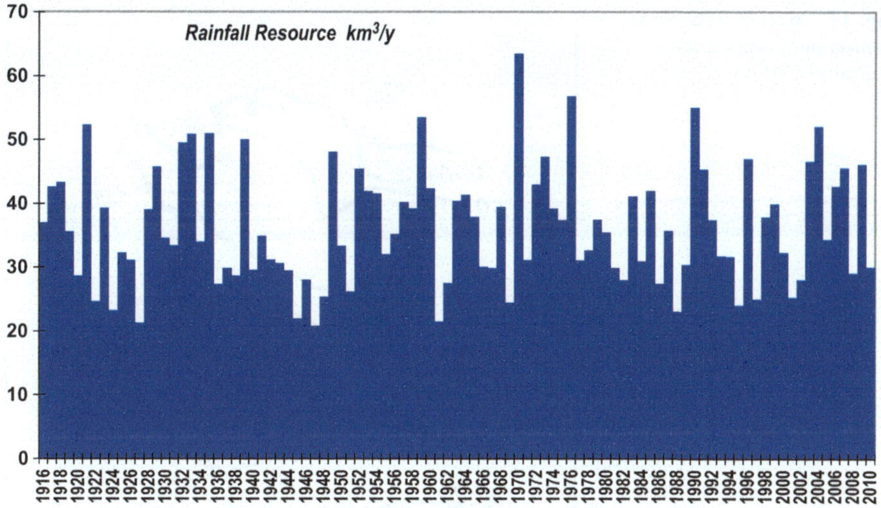

Fig. 4.4 Rainfall water resources of Tunisia; on the period 1916–2010

continental and insular national territory, to an average volume precipitated equal to 36 km^3/year, with a maximum of 64 km^3 (1969–1970) and a minimum of 21 km^3 (1946–1947). The Fig. 4.5 indicates the average isohyets of the country.

4.3.3 *Surface Water Resources*

Surface water resources are the runoff concentrated in streams as flood discharge and base flow.

For the three natural regions of the country, contributions to surface water are as follows: (i) The North, which covers 28% of the country area, provides important and regular surface water estimated at 2.2 km^3/year, or 82% of the total; (ii) The Centre covers 28% in area and brings irregular resources estimated at 0.28 km^3/year or 10% of the total; (iii) The hyper-arid south, covering an area of 44% provides only 0.22 km^3/year or 8% of the total. The national surface water resources are estimated at 2.7 km^3/year in average year, mobilized by large dams, dams and hill lakes, spreading bunds floods, groundwater recharge structures, erosion control bench, cisterns.

The surface water resource presents a very high inter-annual variability, with a minimum of 0.78 km^3/year observed in 1993–94 and a maximum of 11 km^3/year in 1969–70. The max/min annual contribution varies between 9 in North to 180 in the South. The basic average flow is estimated at 0.415 km^3/year, and in case of widespread drought, the absolute minimum flow on the whole country is reduced to 0.150 km^3/year, (Frigui 2005). Furthermore, the quality of surface water greatly

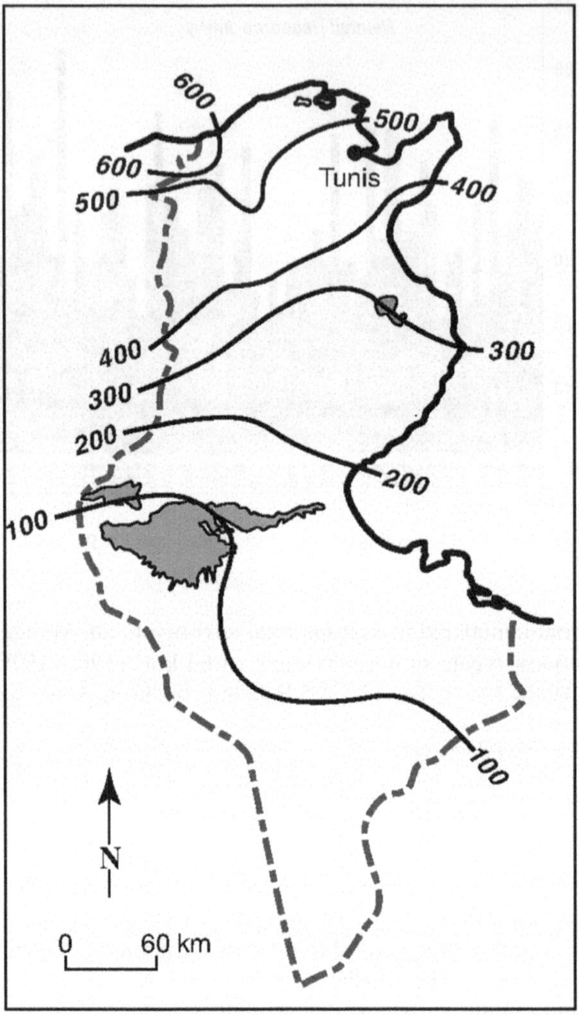

Fig. 4.5 Average isohyets of Tunisia (mm/year). *Source* Benzarti (2003)

varies on the regions: while in the North, 82% of surface water salinity is lower than 1.5 g/L, this proportion is 48% in the Center and only 3% in the South, (Kallel 1994).

Transboundary surface water

The major transboundary water systems are within the following three hydrographic units:

(i) The far North including Oueds Barbara and Melilla, develops in Tunisia on a 240 km^2 basin, crosses the Algerian border and joins the Annaba Wadi Kebir, carrying an average flow of 180 million m^3/year;

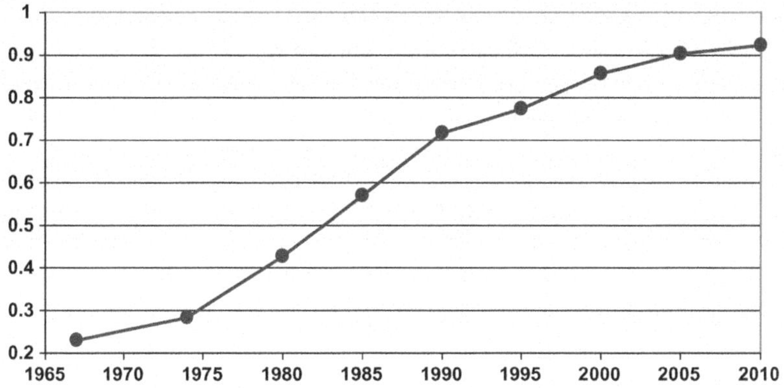

Fig. 4.6 Groundwater withdrawals index on 1968–2010

(ii) The Mellègue-Medjerda watershed, originates and develops in Algeria on 7500 km^2, cross the border with an average flow of 250 million m^3/year;
(iii) The South-Centre, including Oueds Safsaf and Kebir, brings 25 million m^3/year from Algeria.

For 25 years, the two countries, respectively Algeria and Tunisia, unilaterally and without prior agreement, opted for an almost total mobilization of surface water flowing over their own territories: Algeria, with the construction of a series of dams on Medjerda and Mellègue, Tunisia by dams in the far North.

4.3.4 *Exploitable Groundwater Resources*

Exploitable groundwater resources of Tunisia are assessed by DGRE to 2.17 km^3/year, which consist of:

(i) 0.745 km^3/year for the renewable resources of all water table aquifers, according to the 2005 shallow aquifers exploitation yearbook (DGRE 2005).
(ii) 1.42 km^3/year represent the exploitable resources of all the deep aquifers, according to the 2007 deep aquifers exploitation yearbook (DGRE 2007a).

The abstracted volumes are estimated at 2 km^3/year: 0.81 on shallow aquifers, 1.19 on deep aquifers, which represent a Withdrawals Index [WI = withdrawals/resources] equal to 92%. Figure 4.6 shows how this index has changed over the last 40 years, with a steady increase from 1975 to 2000, then a slight stabilization between 2005 and 2010, but over a too short period to be meaningful.

The value of this indicator at the national level masks disparities at two levels. First between shallow and deep aquifers: while in deep aquifers WI is 83%, it exceeds 108% with regard to shallow aquifers. Then, at the regional level: for the 24 governorates, five [Nabeul, Kairouan, Tozeur, Kebili and Gabes] account for almost 50%

of national resources and 50% of withdrawals. The highest overexploitation rates (WI > 100%) are observed in Nabeul, Kairouan and Kebili.

The above observations are based on the assessment of withdrawals and exploitable resources, but each of these balance terms is subject to considerable uncertainty. It is therefore necessary to confirm these estimates and trends by recording measurable parameters, most accessible and most significant for quantitative resource, that is piezometric level. In this regard, regular observations of groundwater levels began in Tunisia more than sixty years ago. Today, the national piezometric network is composed of 3800 monitoring wells (Horriche and Besbes 2006) including a series of sometimes lengthy measures of fifty years.

4.3.5 Renewable and Weakly-Renewable Resources: Return on Exploitable Groundwater Resources

In quantitative terms and according to their recharge process, aquifers of Tunisia may, in a preliminary view and rather reductive mode, be classified in two broad categories: (i) renewable aquifers, (ii) weakly renewable aquifers.

The renewable groundwater resources result from infiltration of precipitation, and recharge by wadis runoff that reaches groundwater, defined in average hydrological year. Estimation of renewable resources, and groundwater recharge, requires long and rigorous protocols, used in scientific research work on specific aquifers. There is still no major national research program dedicated to this issue.

As for weakly-renewable groundwater resources, managing their withdrawals uses simultaneously the concepts of flow and stock. The extracted volumes come from both (i) the aquifer renewal flow, (ii) the stock of reserves accumulated over periods of geological time. This certainly characterizes all aquifers (where abstraction takes necessarily a part but in unequal proportion on the geological reserves) but especially the large sedimentary basins aquifers, in which the part abstracted on reserves can become dominant and decisive.

In large aquifers, beyond the average recharge rate, the "exploitable" part, taken on reserves, should be expected to produce an environmental impact (aquifer drawdowns and salinity changes) considered "acceptable" by the national community and to future generations. In Tunisia, these considerations relate exclusively to the South regions: it is the case of the North-Western Sahara Aquifer System, the NWSAS, including deep aquifers of the Continental Intercalary (IC) and the Terminal Complex (TC), and to some extent that of Djeffara plain, the coastal plain of Medenine-Gabes.

The aquifers of southern Tunisia were subject of extensive recent hydrogeological studies and numerical modelling: the model of Djeffara (OSS 2006) and the NWSAS model (OSS 2003). The results of the Tunisian part of the NWSAS, calculated on the century 1950–2050 (Table 4.1) indicates that the system input flow is currently divided into three equal parts: (i) the local aquifer recharge, (ii) the transboundary

Table 4.1 Balance of the Tunisian part of the North-Western Sahara Aquifer System, respectively in 1950, 2000 and 2050

	1950	2000	2050
Inputs Mm^3/year			
Local recharge	210	210	210
Flow across borders (Algeria & Libya)	180	205	182
Reserves depletion		215	321
Outputs Mm^3/year			
Drainage by Chotts	185	40	0
Tunisian Outlet (Chott Fejej)	105	49	3
Springs	60	1	0
Pumping	40	540	710

inflow, (iii) abstractions on geological reserves, and these proportions should evolve slowly but keep similar orders of magnitude in the future.

On the NWSAS model, the scenario whose impacts were considered acceptable for 2050, in Tunisia, Algeria and Libya, is the one called CI8-CT5, defined together by the three countries (OSS 2003). In this scenario, Tunisian abstractions are brought from 540 million m^3/year, their current level, to 710 million m^3/year, or 150 and 560 for the IC and TC respectively. In this scenario: (i) in the IC, additional withdrawals are essentially delocalized in the extreme south, the rest is scattered ensuring maintenance of flowing levels in Djérid and Nefzaoua; (ii) in the TC, additional withdrawals are minimal around the Chott Jerid, a very high-risk area; the rest is re-deployed in remote areas southern Nefzaoua (Tembain, El Ouaar) and could constitute in the medium term an alternative solution for pumping and transfer to contribute to any future safeguard of Nefzaoua oasis.

On the Djeffara Model, which integrates deep and shallow aquifers, the system recharge is estimated at 110 million m^3/year, and the acceptable scenario brings the system abstractions from 150 million m^3/year, which is the current level, to 210 million m^3/year in the long term. With this scenario, the risk of marine invasion in 2050 is null, and drawdowns are small and localized to the pumping fields.

Taking into account the local aquifers resources in Nefzaoua and Djerid, estimated at 38 million m^3/year, both models will have helped define to 958 million m^3/year all exploitable water resources in southern Tunisia (710+210+38), among which 358 million m^3/year, nearly 38%, are the part strictly renewable "in situ". The remaining 600 million m^3/year comes at a rate of 66% from withdrawals on geological reserves, which manifests continuous drawdowns, and 33% from neighboring countries, the latter part being partly taken from the reserves and the remainder renewed and supplied outside the national territory.

On the rest of Tunisia, which covers parts of North and Centre, groundwater is entirely renewable and exploitable resources are consistent with those identified in

Table 4.2 Classes of groundwater withdrawals by salinity, according to DGRE 2005 (shallow aquifers), 2007a (deep), and DGRE databases

Salt concentration, g/L	C<1.5	1.5–3	3–4	4–5	C>5	Total
Abstractions on shallow aquifers Mm^3/year	52	302	215	60	179	808
Abstractions on shallow aquifers (%)	6	37	27	7	22	
Abstractions on deep aquifers Mm^3/year	302	653	83	102	51	1191
Abstractions on deep aquifers (%)	25	55	7	9	4	
Total aquifers Abstractions Mm^3/year	354	955	298	162	230	1999
Total aquifers Abstractions (%)	18	48	15	8	12	

the latest DGRE yearbooks, which define (i) the Northern renewable resources to 676 million m^3/year (ii) the Centre's renewable resources to 699 million m^3/year.

Thus, and to conclude on this issue:

(i) **All renewable groundwater internal resources of Tunisia are estimated at**: $676+699+358=1733$ million m^3/year
(ii) **All long term exploitable groundwater resources, regardless of quality criteria are evaluated at**: $1733+600=2333$ million m^3/year

including about 200 million m^3/year from outside the country's borders.

4.3.6 Exploitable Resources, Taking Account of the Salinity Criterion

Qualitatively, the Tunisian geological deposit, dominated by Tertiary formations sometimes rich in clay-gypsum elements, endowed Tunisia groundwater with land-based salts concentrations, sometimes significant. If one measures the acceptability level of water resources uses in relation to the threshold of 1.5 g/L, which is the objective of quality supply for two vital sectors such as drinking water and irrigation, and relative to the threshold of 4 g/L which is an irrigation usage limit under certain conditions, one can consider that:

(i) Only 18% of groundwater has salinity less than 1.5 g/L,
(ii) 63% of groundwater has salinity between 1.5 and 4 g/L,
(iii) 19% of groundwater [29% of shallow aquifers, 13% of deep aquifers] has salinity higher than 4 g/L, which is classified as brackish water.

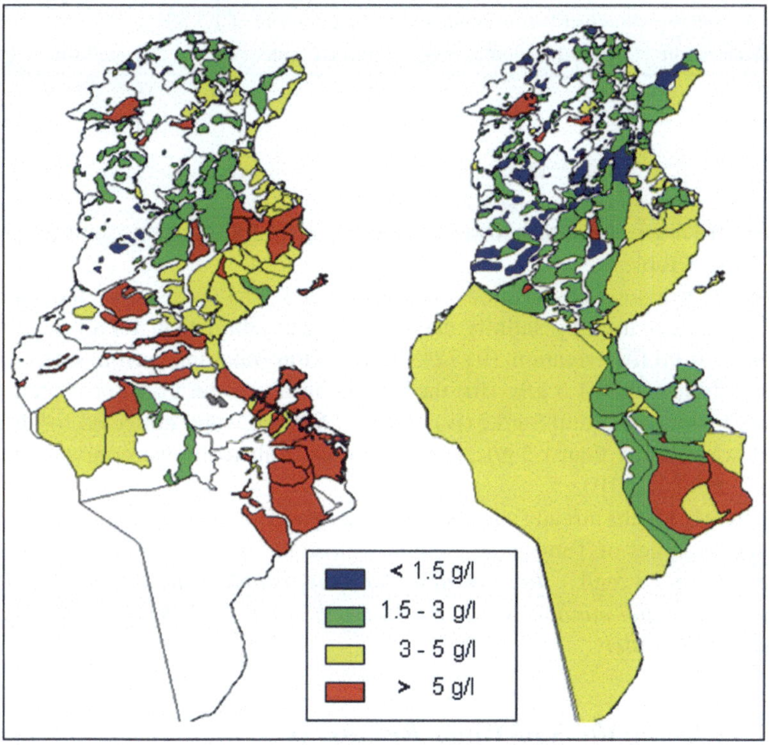

Fig. 4.7 Salinities in shallow and deep aquifers; (*source* DGRE databases)

(iv) Finally, 81% of groundwater are actually exploitable in various areas and in a sustainable way for irrigation, among which only 18% is exploitable for drinking water.

Water salinity distribution also depends on the aquifer systems configurations: Table 4.2 and Fig. 4.7 show how this distribution is much more favorable for deep aquifers, while shallow aquifers have a generally poorer quality.

Applied to exploitable resources, we previously estimated at 2333 million m^3/year without taking into account the quality criterion, the ratios of Table 4.2 suggest to eliminate 20% of underground resources classified as brackish, near 390 million m^3/year the use of which in the long term involves number of risks, including for irrigation.

This would reduce the **estimate of exploitable groundwater resources, with acceptable quality, to the value of 1875 million m^3/year**.

Concentrated in well rain-fed northern regions, surface water have considerably more favorable qualities with 72% having salinities lower than 1.5 g/L. So that, as a national average, the whole water resources are almost equally divided over the

Table 4.3 Water quality distribution compared to the threshold of 1.5 g/L

Salt concentration	Total resources (%)	Surface water (%)	Groundwater (%)
<1.5 g/L	49	72	16
>1.5 g/L	51	28	84

threshold of 1.5 g/L: 49% with lower concentrations, 51% with concentrations higher than 1.5 g/L. (Table 4.3).

Based on previous observations, and on other DGRE yearbooks and estimates for deep aquifers uses by salinity classes, one can admit that: (i) all the shallow aquifers are used for irrigation, (ii) 72% of irrigation water taken from surface water have salinity less than 1.5 g/L, (iii) the rest of surface water used for irrigation has a salinity between 1.5 and 3 g/L, (iv) 100% of surface water collected for drinking have a lower salinity than 1.5 g/L, (v) surface water abstractions represent a total of 780 million m^3 in 2010.

These early results already suggest a critical remark, somewhat paradoxical: while the water resources of Tunisia are generally close to 50% of good physicochemical quality (salinity < 1.5 g/L), ***the current management mode****, which favors exploitation of groundwater and minimizes water surface abstractions,* ***allows using only 35% of good quality water***.

4.3.7 *Defining the Soil Water Resources*

Soil water resource is defined as the infiltrated part of rainfall temporarily stored in the ground, which is available either for direct evaporation in the case of bare soil or for plant Evapotranspiration in case of a covered soil. It corresponds to the concept of useful reserve (UR) of agronomists, used in conceptual hydrological models. In theory, the entire stock of ground water within the UR can be used by plants in the case of a very dense cover like a forest or a lawn.

The soils water resources in Tunisia concern arable land (5 million ha), rangelands (5 million ha) and forests (nearly one million ha) on the area of which is integrated the rainfall resource. Precise evaluation of this resource throughout Tunisia is complex. A number of hydrological estimates limited to arable land resource have been proposed around 12-15 km^3/year on average hydrological year. An indirect estimate, computing the water equivalent of national agricultural production (Besbes et al. 2002), provides for a default value (concerns only the quantities used by rainfed and irrigated cultures), at nearly 7 km^3/year. In the study of actual Evapotranspiration (WaterWatch 2008), a more extensive evaluation (for all potential arable land) provides for a much higher estimate to about 13 km^3/year, to which should be added the rangelands and forests contributions.

4.4 The Water Resources Mobilization

4.4.1 Large Dams, Hill Lakes and Hill Dams

In 2010, Tunisia had 30 large dams, whose useful capacity is estimated at 2.75 km^3, to regularize a total contribution of 1.7 km^3/year (average contribution observed over the last 20 years is 1.65 km^3/year). Added to this are 225 small dams (hill dams) and 750 small lakes (hill lakes), mobilizing respectively 0.15 and 0.05 km^3/year. This entire water infrastructure mobilizes 1.9 km^3/year, or 70.4% of all surface water resources.

Surface water resources (average annual flows) are estimated at 2.7 km^3/year. This value is validated by the analysis of a series of 45 annual values (1960–2005) observed throughout the country (Frigui and Touzi 2009). This analysis confirms the great variability of surface flows: to have the average flow every two years, the storage capacity must provide, compared to the allocations, an additional reserve of 400 million m^3, and even more when one assigns greater security, taking into account evaporation in reservoirs.

4.4.2 Shallow Aquifers and Deep Aquifers

The shallow aquifers resources represent 50% of renewable groundwater resources in Tunisia: with data and knowledge progress, the estimated overall volume of renewable available resource has increased from 486 million m^3/year in 1980 to 746 million m^3/year in 2005, and total withdrawals increased from 395 million m^3/year in 1980 (with 23,000 pumped wells) to 810 million m^3/year in 2010 (with more than 100,000 pumped wells).

The estimate of exploitable deep aquifers resources has increased from 1000 million m^3/year in 1980 to 1420 million m^3/year in 2010; with abstractions growing from 530 million m^3/year to 1190 million m^3/year during the same period.

The number of boreholes drilled in Tunisia from 1991 to 2010 amounts to 9300 boreholes, nearly 470/year on average. These holes are divided into 2900 public wells and 6400 regularly allowed private wells.

In space and time, one can observe quite significant trends:

(i) The number of private wells, ridiculous over the 80s, exceeds that of public wells in 1996 and permanently since 2000 (Fig. 4.8). This trend has been accentuated, reaching 92% of total drillings in 2010. This observation applies only to private wells regularly allowed.

(ii) By governorate, the boreholes distribution is uneven: three governorates (Nabeul, Ben Arous and Kasserine) account for over 50% of the total number of wells.

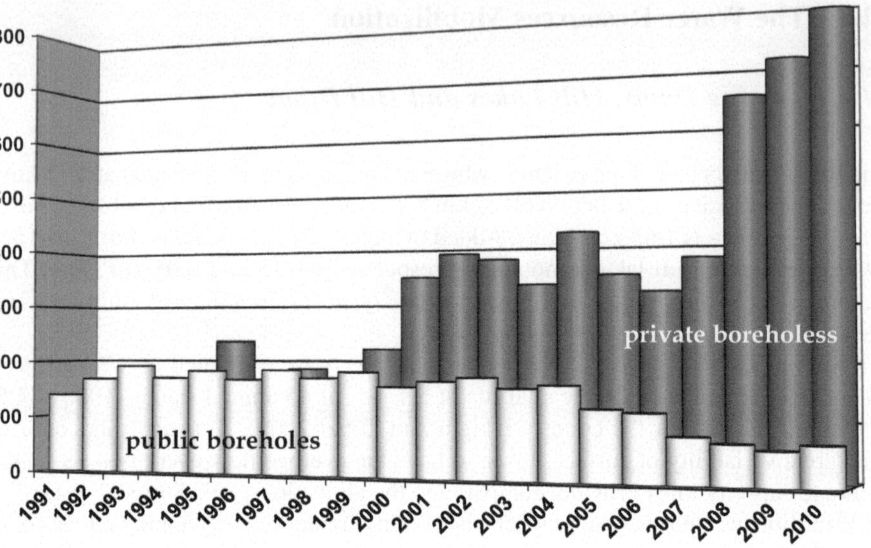

Fig. 4.8 Number of public and private boreholes, drilled from 1991 to 2010 (data from DGRE databases)

4.4.3 Evapotranspiration and the Green Water Resource

The Evapotranspiration (ET), sum of water released into the atmosphere by evaporation from the soil surface and by transpiration from plants, is an essential component of the water cycle. In a given region and throughout the season or year, the quantities of water mobilized by humans and ecosystems generally represent the difference between the cumulative rainfall and evapotranspiration (ET). But the accurate determination of these two quantities is facing the same difficulties, because of the very high spatial and temporal heterogeneities due to variations in geographical and climatic conditions. Remotely sensed data can provide continuous observations of rainfall field and ET. To this end, the Moderate Resolution Imaging Spectroradiometer (MODIS) onboard of the TERRA and AQUA satellites provides accurate information on vegetation and the energy balance at the soil surface, allowing a reliable modeling of ET (Mu et al. 2007). This model is validated by a global network of ground measurements stations, and provides the global maps of Evapotranspiration, rainfall, and PET which represents the total flow, all integrated on agricultural land areas.

In Tunisia, the knowledge of the annual ET balance allows to specify, on the amounts of water applied from precipitation and irrigation, what are the volumes actually consumed and discharged into the atmosphere. WaterWatch (2008) uses TRMM signals (Tropical Rainfall Measuring Mission, a common mission of Japan Aerospace Agency JAXA and the NASA), calibrated on the Tunisian National Meteorological Institute (INM) network to generate monthly rainfall throughout cultivated areas. Over the two chosen reference years, 2000–2001 and 2006–2007, modeling

Table 4.4 WaterWatch results for the two reference years: values in km^3/year

	Hydrologic year	Irrigated area	Rain-fed area	Total agriculture
		3950 km^2	41,000 km^2	44,950 km^2
Rainfall supply, km^3	2000–2001	1.02	8.9	9.92
	2006–2007	1.6	14.86	16.46
Supply to the plot, surface and groundwater, km^3	2000–2001	0.71		0.71
	2006–2007	0.56		0.56
Total water supply, km^3	2000–2001	1.73	8.9	10.63
	2006–2007	2.15	14.86	17.01
Evapotranspiration of irrigated areas, km^3	2000–2001	1.69		1.69
	2006–2007	1.85		1.85
Evapotranspiration of rain-fed areas, km^3			9.56	9.56
			12.38	12.38
Total evapotranspiration of cultivated area, km^3	2000–2001	1.69	9.56	11.25
	2006–2007	1.85	12.38	14.23
Water Balance, km^3	2000–2001	0.03	−0.66	−0.63
	2006–2007	0.31	2.49	2.8

ET provides map of the actual Evapotranspiration, which is compared to the rainfall map, (WaterWatch 2008). Figures 4.9, 4.10 and Table 4.4 present the overall results.

The results of the study can be summarized in the water balance presented on last rows of Table 4.4. For the year 2006–2007, a wet year, the balance is positive: actual Evapotranspiration is less than the sum of the inputs (rain and irrigation); the excess will constitute water resources: surface runoff and deep infiltration. As for 2000–2001, a dry year, it shows a negative balance: the deficit is taken from the soil water reserve accumulated in the previous year. Note that the irrigated areas balance is zero in dry years: irrigation dose is just sufficient to offset ET.

4.4.4 *Surface Water Mobilization and Soil and Water Conservation Techniques*

It is conventionally recognized that on 2.7 km^3/year of surface water, a portion, about 0.6 km^3/year of weakly concentrated runoff, can be mobilized only by Soil

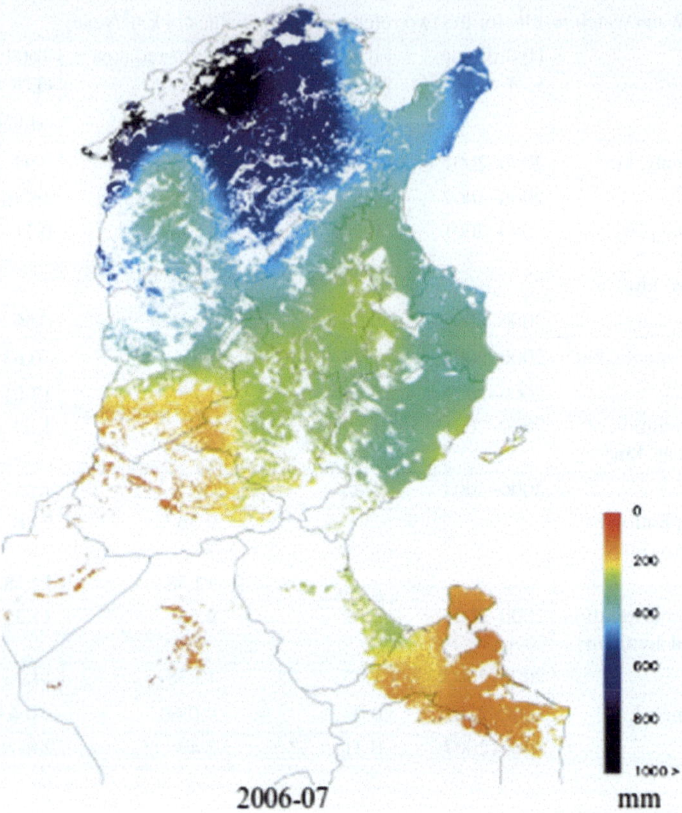

Fig. 4.9 Precipitations on agricultural areas in 2006–2007 (according to WaterWatch 2008)

and Water Conservation (SWC) techniques, the most common are the rainfall collection techniques or traditional terraces, mechanized from decades as DRS benches (Defense and Restoration of Soils) or SWC (Soil and Water Conservation). Over the past 40 years, these mechanized erosion control benches covered in Tunisia nearly a million hectares of cropland and rangelands. These arrangements caused spectacular rangelands regeneration. But what has been the impact of these works on water balance? If these works, properly made, are unquestionably effective in limiting erosion, they are not always favorable for the water balance as they surely decrease runoff and increase effective infiltration if they are situated on aquifer recharge areas, but this happens exceptionally. These structures are in any way known to increase a little more Evapotranspiration, favoring accumulation of soil water resources and production of green water.

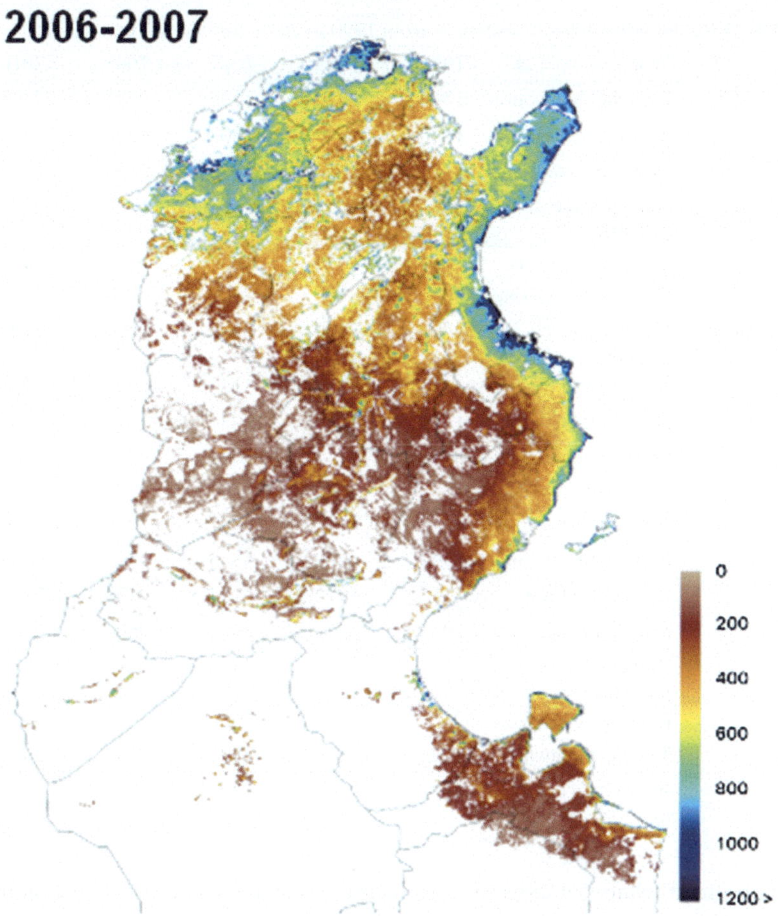

Fig. 4.10 Actual Evapotranspiration on cultivated areas in 2006–2007 (according to WaterWatch 2018)

4.5 The Status of Water Uses

Agricultural uses, drinking water, industrial, touristic use, and recycled waters, have been checked in Chap. 3.

Environmental demand and water balance for uses

Realization of the different water development programs has led to a sharp artificially water cycle. This results in a reduction in hydro-systems recharge, with consequences on continental and coastal aquatic ecosystems. In these circumstances, the environment protection requires systematic consideration of water resource allocation

Table 4.5 Different uses of water resources, from 1990 to 2010, million m^3/year

	Year	1990	1995	2000	2005	2010
Groundwater	Exploitable resource	1840	1929	2135	2155	2165
	Agricultural use	1330	1440	1608	1660	1705
	Domestic use	148	168	175	213	250
	Non connected industries	56	65	74	61	65
	Total withdrawals	1534	1673	1857	1934	2020
	% resource exploited	83%	87%	87%	90%	93%
Surface water	Mobilizable resource	2700	2700	2700	2700	2700
	Agricultural use	245	532	516	478	480
	Domestic use	143	155	193	237	300
	Environmental uses	0	0	30	60	100
	Total withdrawals	388	687	739	775	880
	% resource exploited	14%	25%	27%	29%	33%
Total	Exploitable resource	4540	4629	4835	4855	4865
	Agricultural use	1575	1972	2124	2138	2185
	Domestic use	291	323	368	450	550
	Non connected Industries	56	65	74	61	65
	Environmental uses	0	0	30	60	100
	Total withdrawals	1922	2360	2596	2709	2900
	% resource exploited	42%	51%	54%	56%	60%

for environment: water releases to lakes feeding, wetlands and natural groundwater recharge.

The characteristics of the Tunisian climate also require preserving a certain volume stored in the reservoirs for later use, which can represent nearly a third of mobilized surface water. Indeed, when mobilized by regularizing a large part or even the whole of the surface water resources, the annual volume actually used should always be less than the volume regulated to ensure that the storage reservoirs function not only for seasonal needs but also for inter-annual regulation.

Table 4.5 summarizes all the water allocations between 1990 and 2010, classified by origin of the resource.

Reading this table suggests that groundwater is exploited today at a rate of 93%, while surface water would be at only 29% if we consider only traditionally recognized water uses (domestic, industry, irrigation), but if we extend the exercise to environmental uses, detailed as:

(i) artificial wetlands feeding, about 100 million m^3/year (currently limited to Ichkeul system);

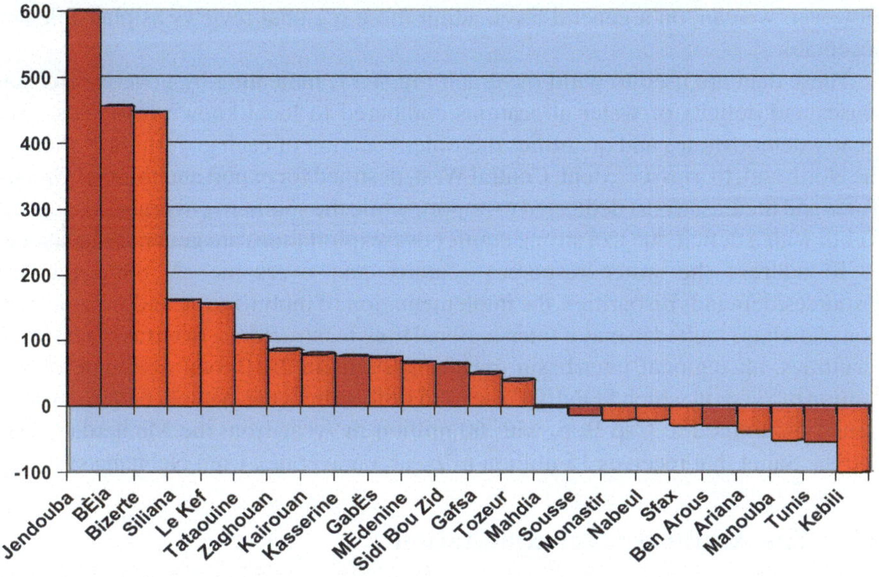

Fig. 4.11 Regional water resource minus allocation assessments by governorate, million m^3/year, reference year 2010

(ii) artificial groundwater recharge limited to overexploited aquifers, estimated at 50 million m^3/year on average;
evaporation in reservoirs, estimated at 150 million m^3/year;

In this case, "effective" abstraction from surface water established in 2010 would increase from 780 million m^3/year (Table 4.5) to a total of 1080 million m^3/year, representing an effective rate of 40%. Such a description, however, must take into account the fact that large water projects (drinking water and irrigation) gradually exploit the resources reserved for them and the momentary under exploitation of surface water is generally doomed to be obliterated in the medium term.

4.6 Water Resources and Uses: The Regional Imbalances and Major Transfers

By major region and by governorate, regional balance results from crossing and treatment of multiple data sources, which are not always synchronous or homogenous: DGRE yearbook of deep aquifers, DGRE yearbook of shallow aquifers, SONEDE annual Statistical Report, several notes from DGGREE. This diversity of sources leads to carefully consider the details results achieved at the end of this exercise.

However, we can, on a general level, admit these regional reviews as plausible and acceptable.

These data are used to build the graph Fig. 4.11, indicating by governorate, surpluses and deficits of water allocations compared to local known resources. This clearly demonstrates and quantifies the water vocation of each governorate: those in the North and, to a lesser extent, Central West, destined for export and those of Greater Tunis and the East Coast dedicated to import, while the southern governorates (except Kebili with a deficit due to a strong aquifer overexploitation) are generally balanced.

To address the water resources scarcity and overcome all these regional resources/demands disparities, the implementation of mobilization and water allocation plans has resulted in major inter-regional transfer structures. In terms of conveyed quantities, all regional inter-basin transfer systems in 2010 total a volume of 400 million m^3/year in which SONEDE uses 325 million m^3/year, namely 81%. Irrigation transfer is limited to Cap Bon, with 60 million m^3/year from the Medjerda Canal, and the Sahel that receives 15 million m^3/year, on average, from Nebhana system.

4.7 The Aquifers Overexploitation

4.7.1 An Advanced State of Overdrafts

Over the past three decades, the investment effort in the hydro agricultural sector has been accompanied by encouraging increased pumping with wells electrification, grants for well construction and pumping equipment. These measures have led to a rapid and significant increase in withdrawals with: (i) creation of new boreholes and wells, (ii) acquisition of pumping equipment increasingly powerful. Thus, withdrawals on aquifers more than doubled on a 25 years period: from 0.9 to 1.95 km^3/year between 1980 and 2005. These quantities are close to or exceed the exploitable groundwater limits of the country.

Despite the significant economic and social impact due to the extension of a very rich irrigation activity, this has led to significant aquifers drawdowns (Fig. 4.12), and a large proportion of aquifers are now labeled overexploited by DGRE (2007a, b): almost 55 shallow and 30 deep aquifers. These aquifers have surplus withdrawals estimated at 400 million m^3/year for the whole Tunisia. Kebili governorate, Nabeul, Kairouan and Sidi Bou Zid abstractions count more than 75% of these water deficits. Furthermore, a specific estimate of abstractions on geological reserves in south weakly renewable aquifers (Terminal Complex and Intercalary Continental) gives nearly 200 million m^3/year, or 50% of all groundwater overexploitation rates of Tunisia.

Many aquifers are declared "safeguard perimeter" by decree. But these measures could not stop wells proliferation. Certainly, no subsidy is granted for new well construction in these areas, contributing to slowing their progress. Farmers do not ask permission for the creation of new wells and many people build their wells without any subsidies. Moreover, and due to the regular lowering of piezometric levels, farmers conduct deepening their wells by hand probes that lead beyond the limit of 50 m, the regulatory depth limit beyond which a prior authorization is required.

Overexploitation reduces the geological reserves of aquifers; the continual lowering of overexploited aquifers levels is then likely to eventually cause a total depletion of the most vulnerable aquifers. Figure 4.13 shows respectively: (i) the shallow aquifers limits, (ii) the currently overexploited shallow aquifers, (iii) shallow aquifers that will be fully depleted before the end of the XXIth century at the current withdrawals rate.

Groundwater is a distributed resource, naturally shared: groundwater is like a network operated collectively and interactively by hundreds or thousands of users. To prevent abuse, the water authority must promote negotiation processes between users. Transposed to the regional scale of transboundary aquifers management, this principle resulted to set up a consultation mechanism to rationalize the exploitation, with Algeria and Libya, of the shared large southern Saharan aquifers. Locally, participatory management of overexploited groundwater was attempted: to preserve the coastal aquifer heavily used north of Gabes, nearly 300 farmers have created the Bsissi groundwater development, operation and monitoring Group. This association got that illegal boreholes are resealed and the digging of new wells is stopped. The Group is responsible for controlling the allocated pumping rates and monitors the piezometric levels. This positive experience however, remains fragile and unique in Tunisia. In the perspective of a reproduction of that community management experience, the Ministry of agriculture tries to change the regulatory and legislative framework user associations to make it compatible with groundwater management.

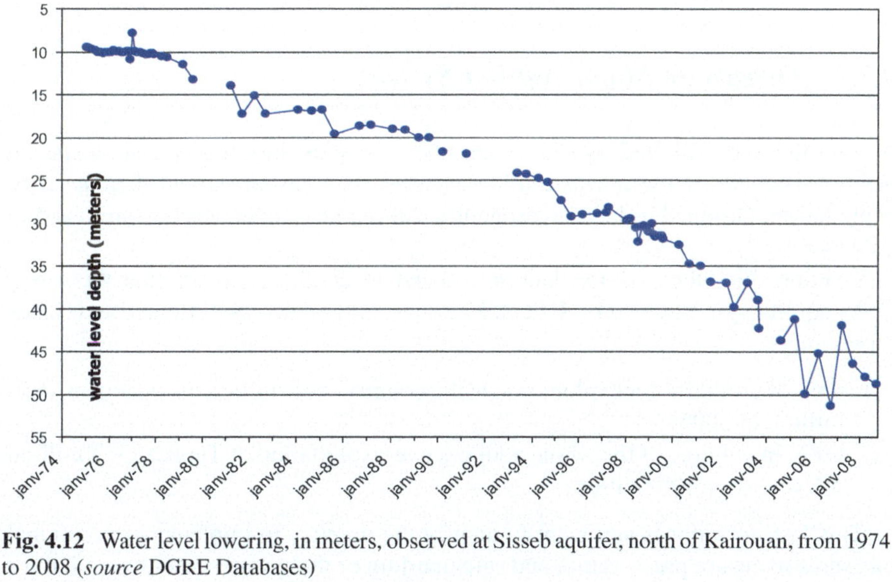

Fig. 4.12 Water level lowering, in meters, observed at Sisseb aquifer, north of Kairouan, from 1974 to 2008 (*source* DGRE Databases)

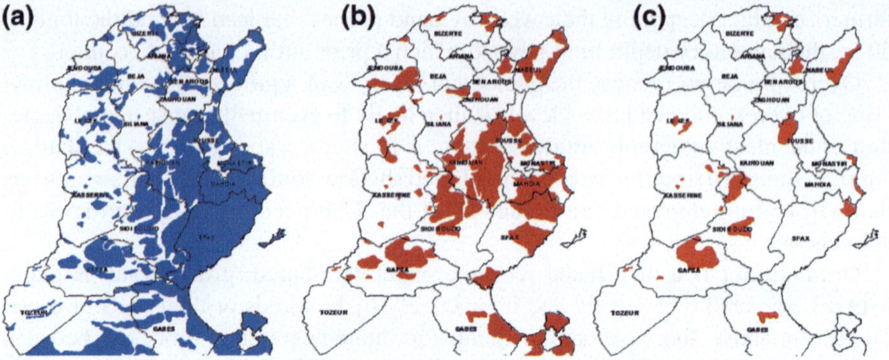

Fig. 4.13 Status and future of shallow aquifers **a** shallow aquifers extension, **b** over-exploited shallow aquifers, **c** those that will be fully depleted before the end of the XXIth century

Table 4.6 Overexploitation of major aquifer regions in 2005 (*source* DGRE 2007a, b)

Aquifer region	Resources Mm^3/year	Withdrawals Mm^3/year	Overdraft Mm^3/year	Overdraft index (%)
Kebili	173	341	168	197
Nabeul	147	226	79	154
Kairouan	73	120	47	164
Kasserine-Sidi BouZid	65	113	48	174
Gabès	50	65	15	130
Total	508	865	357	170

4.7.2 Threats on Major Aquifer Systems

Among the overexploited aquifers, a number occupies, in terms of risk, a special position because of the intrinsic importance of their resources and their level of exploitation. Grouped in large hydrogeological regions, these aquifers are listed in Table 4.6.

Reading this table, whose data are related to 2005, it appears that these five major aquifer systems, Kebili, Nabeul, Kairouan, Kasserine—Sidi Bouzid and Gabes represent:

(i) 45% of groundwater exploitation in the country (865 million m^3/year over 1950 million in 2005)
(ii) 90%, in volume, of the whole aquifers overexploitation of Tunisia (357 million m^3/year over 400 million).

This last indicator is sufficient to qualify the urgency to implement exceptional measures to ensure preservation and safeguarding of these water resources, vital for the future of the country (Figs. 4.14 and 4.15).

4.8 The Anthropogenic Water Cycle and the Whole Water Balance of Tunisia

4.8.1 Mobilization, Use, Consumption: The Anthropogenic Water Cycle

The compilation of a lot of data related to water resources mobilization is presented by the following graph, Fig. 4.17 "anthropogenic conventional water cycle", where one tries to distinguish between mobilized resources, exploitable resources, allocations and abstractions, actually used resources, net consumption, and returns to the receiving environment.

The ***Mobilized Resources*** are quantities effectively developed through existing hydraulic structures. These amounts represent an annual average potential equal to 3960 million m^3/year (Basis: 2010)

The ***Exploitable Resources*** differ from mobilized resources by the quantities actually mobilized [available in reservoirs dams for example] but which does not carry out immediate use to reflect forecasted allocations to new projects already decided when creating the dam. This transitional reserve may be estimated at 600 million m^3/year.

The ***Withdrawals*** are determined by resource type as follows:

In reservoirs dams, the useful exploitable volume [930 million m^3/year] is composed by:

Fig. 4.14 Illegal drilling, 120 m depth with a homemade rig; Bled Sisseb north of Kairouan, 21/01/2013. The number of unauthorized wells drilled in this area is estimated to more than five hundred, and the water table has dropped by 40 m in 30 years

Fig. 4.15 New palm around Kebili; unlawful extension of unauthorized drilling; 26/6/2007. The number of illegal inventoried boreholes in Kebili, in 2008, was higher than 3000

- 150 million m^3/year which form evaporation in water reservoirs,
- 100 million m^3/year which are reserved for the regulation of wetlands (Ichkeul)
- 300 million m^3/year collected by SONEDE,
- 600 million m^3/year reserved to irrigate the "Public Irrigated Perimeters",
- 50 million m^3/year (average) assigned to groundwater recharge,
- 140 million m^3/year are not explicitly assigned and represent closing term of dams water balance, which incorporates uncertainties on the balance sheet terms, both on inputs (superficial contributions) and outputs (leakage rates, not recorded abstractions).

In shallow aquifers, 820 million m^3/year are abstracted for private farms irrigation (greater than renewable resources because of overuse).

In deep aquifers, the exploited volumes [1200 m^3/year] are distributed as follows:

- 240 million m^3/year collected by SONEDE,
- 60 million m^3/year abstracted by Industries not connected to public networks,
- 900 million m^3/year taken for public and private Irrigation perimeters.

The ***Actually Used Resources***:

On an allocated volume of 3020 million m^3/year, only 2095 are actually used, or nearly 70%; the rest, nearly 925 million m^3/year, is distributed as follows:

- 145 million m^3 /year represent various losses of public networks (drinking water)
- 660 million m^3/year constitute the total losses of the irrigation system,
- 120 million m^3/year represent estimated volume in reservoir dams, reserved for Public Irrigation Perimeters and not used by farmers.

The losses representing all these amounts relative to allocated volumes are temporary or definitive, sometimes incompressible as some network losses; finally some are voluntary and desirable, such as land irrigation needed for salts leaching.

Ultimately, the 2095 million m^3/year actually used are divided, by user type, into:

- 350 million m^3/year for drinking water,
- 100 million m^3/year for wetland ecosystems,
- 105 million m^3/year for industry,
- 1540 million m^3/year for irrigation.

The ***Net Consumptions*** are the "vanished" part of the used resource, that is actually consumed and therefore has definitely left the continental water cycle or the entire water cycle; it is:

- In agriculture, the amount evapotranspired by plants (for the greater part) and those forming part of the biomass of plants and food products. The share of net consumption in agriculture is generally estimated at 66%.
- For drinking water, the net consumption is estimated at 20%
- In industry, the net consumptions are at 20% on average.
- We have also the re-evaporated part of the networks losses, most of which are for irrigation networks.

The general sum of net consumptions amounts to 1580 million m^3/year.

Returns to the receiving environment:

Part of the irrigation and public networks losses go back to the continental natural environment and are therefore likely to return to the natural water cycle. This amount, in first approximation, is not at all negligible as it is in the same orders of magnitude (1480 million m^3/year) than the actual net consumption. Some of these returns to the receiving medium are drained and recycled by mixing in Oued Medjerda, but other parts, sometimes significant, can not be reused due to the water quality degradation after use, in particular when originally irrigation water have significant salinity.

The reconstruction of Anthropogenic Water Cycle necessitated a number of overlaps, simplifications and assumptions, which are presented on the legend below (Box 4.1). This assessment relates to the average data reported to 2010 as reference year.

Box 4.1: Legend of the Anthropogenic Water Cycle

1. Inter annual runoff contributions to reservoirs dams regularization
2. Available on average year in dams reservoirs
3. Shallow aquifers
4. Deep aquifers
5. Evaporation in dams and allocation to wetlands
6. Transfer from SONEDE dams
7. Water from dams devoted to irrigation
8. Not explicitly allocated: closing error or allocated and not used
9. Groundwater artificial Recharge
10. SONEDE Transfer from deep aquifers

11. Groundwater resources allocated to irrigation
12. Unconnected Industries and Tourism
13. Industries and Tourism from SONEDE networks
14. Billed to communities
15. Leaks from networks, including service losses
16. Public Irrigated Perimeters use 80% of the allocated surface water
17. A third of the whole irrigation system losses are re-evaporated
18. Two thirds of the irrigation system losses go back to a receiving medium
19. Quantities actually issued to crops, given the efficiencies
20. Effective consumption of collectivities
21. Effluents from urban networks
22. Effective consumption of industries
23. Industrial Water Releases
24. Actual Evapo-transpiration and integration with biomass
25. Part joining the receiving environment

4.8.2 Becoming of the Rainwater Resources, or the Natural Hydrological Cycle of Tunisia

We are legitimately entitled to ask the question of the fate of rainfall on the country, the famous 36 km^3/year. Actually, this question has met little interest from experts and there are no documentary references to which we can rely. The following estimates, presented on the graph Fig. 4.17, constitute therefore a pure intellectual speculation with uncertain scientific basis and should be considered as a first working hypothesis, plausible after all, for securing orders of magnitude. Below are presented the data on which these first estimates are based, which have led to establish the natural hydrologic cycle of Tunisia:

Table 4.7 summarizes the calculation steps used to arrive at the hydrological cycle general assessment in its current state (reference year 2010). This balance is illustrated by the schematic graph of Fig. 4.17.

How to go from Table 4.7 **to Fig.** 4.17**?**

To clarify the relantionships between Table 4.7, Figs. 4.16 and 4.17, it should be emphasized what follows:

- Due to the endoreism characterizing very large areas of the territory, part of surface and groundwater flow (around 2 km^3/year) is drained and evaporated in downstream humid areas like Chotts and Sebkhas and can not reach the sea; this part is affected to the "humid areas" evaporation box in Fig. 4.17.
  Furthermore, the share of irrigation net consumptive use is estimated at 1.35 km^3/year according to the graph budget Fig. 4.16 (see the last column); this

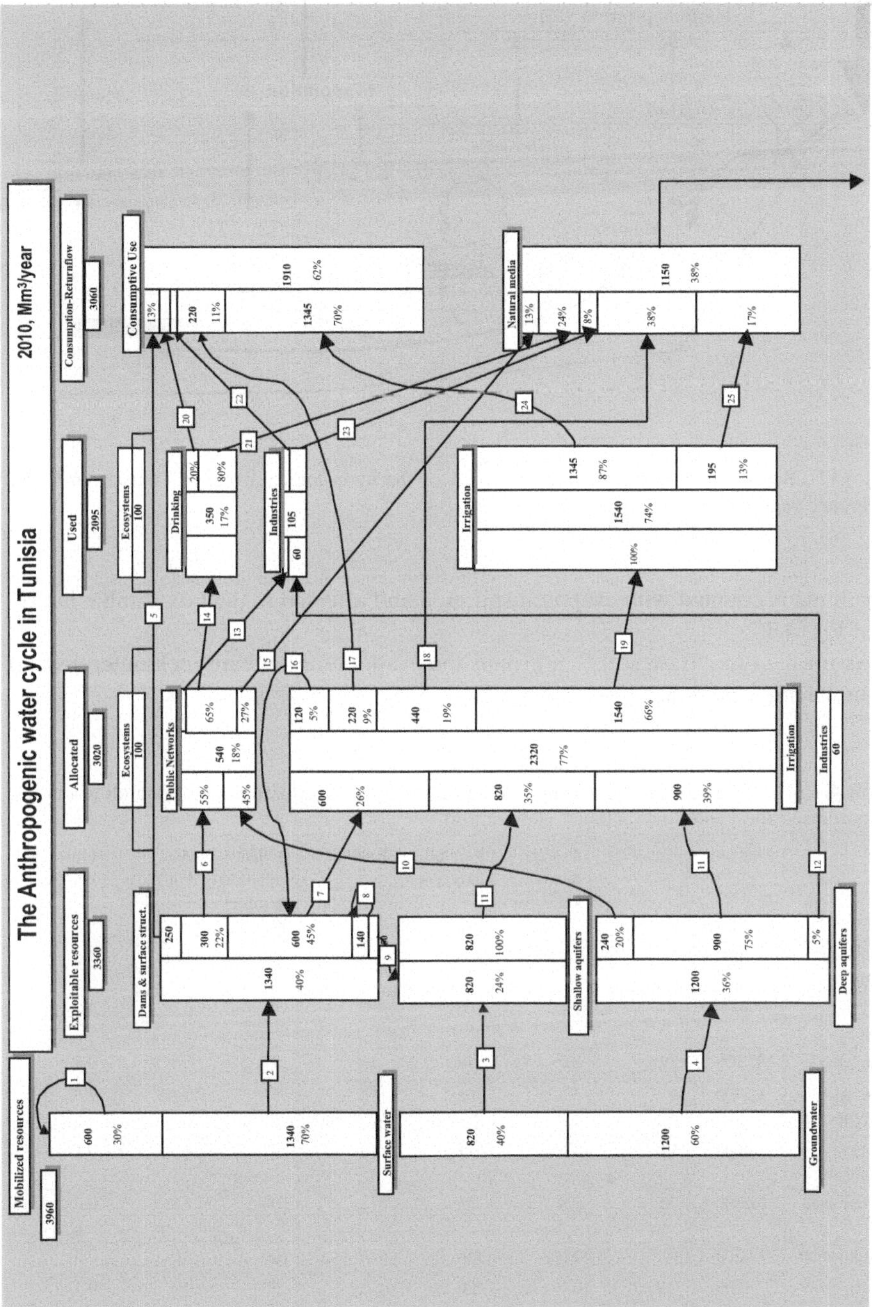

Fig. 4.16 The anthropogenic water cycle in Tunisia, 2010, million m^3 (Mm^3/year)

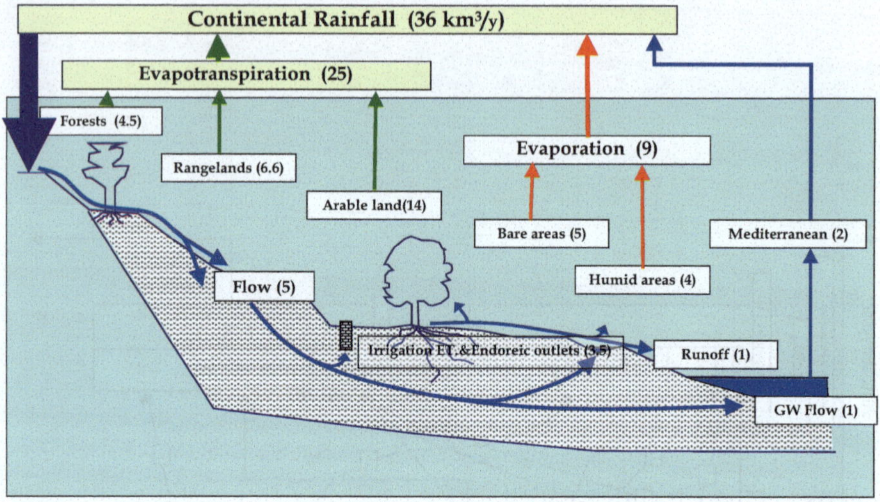

Fig. 4.17 Becoming of the rainwater resources, or the hydrological cycle of Tunisia, km^3/year (reference year 2010)

volume is counted with evapotranspiration and affected to the box "arable lands" of Fig. 4.17.

- As for the box "Bare soils", it groups the evaporation on bare rock outcrops and on Sahara sand dunes respectively.

Table 4.7 The land occupation in Tunisia and the corresponding distribution of rainfall resources, evapotranspiration and flow (authors' elaboration)

	Area km^2	Area (%)	Average rainfall km^3/year	Rainfall mm/year	Rainfall %	Actual evapotranspiration %	AET km^3/year	Flow km^3/year
Rainfed areas	43,000	26	13.2	307	36	88	11.6	1.6
Irrigated area	4000	2	1.2	300	3	75	0.9	0.3
Rangelands	47,000	29	7.0	149	19	94	6.6	0.4
Bare rock outcrops	12,400	8	5.8	468	16	55	3.2	2.6
Forests, scrubland	9000	5	4.6	511	13	97	4.5	0.1
Sahara sand dunes	32,800	20	1.8	55	5	100	1.8	0.0
Humid areas	14,700	9	2.0	136	6	100	2.0	0.0
Urban areas	1500	0.9	0.6	400	2	50	0.3	0.3
Whole country	164,400	100	36	220	100	85	31	5

References

Benzarti Z (2003) La pluviométrie en Tunisie. Analyse des années très pluvieuses; in Eau et environnement, Tunisie et milieux méditerranéens. Lyon: ENS Éditions, 2003. http://books.openedition.org/enseditions/869

Besbes M, Chahed J, Bargaoui Z, Mattoussi MS, Mhiri A (2002) Water future, a new challenge for Tunisia, *int.rep.*Tunisian Institute for Strategic Studies, Tunis

DGRE (2005) Etat de l'exploitation des nappes phréatiques de Tunisie. Ed. DGRE, Tunis

DGRE (2007a) Annuaire de l'exploitation des nappes profondes. Ed. DGRE, Tunis

DGRE (2007b) La surexploitation des nappes; Note inédite présentée pour la préparation du PISEAU; MARH. Ed. DGRE, Tunis

Frigui HL (2005) Rapport National sur la mise en valeur des ressources en eau en Tunisie; Nations Unies; Commission Economique pour l'Afrique; CEA-AN/PUB/EAU/TN

Frigui HL, Touzi S (2009) Caractérisation des événements extrêmes en Tunisie et possibilités de la maîtrise des excédents pour une gestion durable de la ressource; in l'Etat des ressources en eau du Maghreb en 2009; UNESCO, Unesco office in Rabat; 2009

Horriche F, Besbes M (2006) Analyse du Réseau Piézométrique National Tunisien. Rev Sci Eau 19(4):347–363

Kallel R (1994) Caractéristiques morphologiques et hydrologiques du réseau hydrographique Tunisien. DGRE, Tunis

Mu Q, Heinsch FA, Zhao M, Running SW (2007) Development of a global evapotranspiration algorithm based on MODIS and global meteorology data. Remote Sens Environ 111:519–536

OSS (2003) Système Aquifère du Sahara Septentrional, Rapport de synthèse, Observatoire du Sahara et du Sahel, Tunis

OSS (2006) Modèle de la Djeffara tuniso libyenne, Observatoire du Sahara et du Sahel, Tunis

WaterWatch (2008) Water balance and evaluation of water saving investments in Tunisian Agriculture; draft version, 2008. Study under assignment of the World Bank in Washington D.C, USA; S.J. Zwart and W.G.M. Bastiaanssen from WaterWatch in Wageningen

WaterWatch (2018) http://www.waterwatch.nl/projects/africa/tunisia.html, consulted on 6 May 2018

WWAP (2009) The United Nations world water development report 3, case studies volume: facing the challenges. World Water Assessment Programme

Chapter 5
Water Demand Management and Non Conventional Resources

Contents

We first examine aspects relating to conditions of water supply and management, as well as major challenges in key areas of water use, in vital or priority sectors (drinking water needs) and the economic sector that consumes the most (agriculture). Then, possibilities and conditions for use of non conventional water: desalination and reuse of treated wastewater are considered as an extension to demand management policies.

An introductory preamble is devoted to the presentation of general principles of water demand management and the impact of these principles on the general water policies.

M. Besbes et al., *National Water Security*,
https://doi.org/10.1007/978-3-319-75499-4_5

5.1 Principles of Water Demand Management in Tunisia

Water policy based on supply growth has been the traditional response to satisfy the demand increase. Today, this policy is reaching its limits face to increasing economic and ecological barriers in most regions. Aware of these implications, the government is gradually evolving, but still too slowly, toward a water demand management policy, while continuing to develop additional resources. This new direction is intended to become an essential component of integrated water resources management (IWRM). Even if the IWRM is yet in its early stage, it shows for the moment a rather break with the "hard" traditional supply policy, which nevertheless remains necessary under certain conditions to continue developing new offer to supply rural areas, promote industrial activities, and agricultural intensification through irrigation. In areas where large-scale water projects are already in place, improving security of water supply to the different sectors will require interconnections between large reservoirs and large water transfers.

This so-called "weak demand management" stage is characterized by the implementation of technical, economic and regulatory instruments whose main objectives are to reduce losses and misuse in water systems, stabilize the various sectors consumptions, help to diminish pressure on water resources, in particular to reduce or stop unsustainable abstractions (subject to pollution, overexploitation). This first stage of demand management involves, in fact, relatively minor reforms. This transitional solution should ensure better use in each of the sectors taken independently without taking into account the competitive relationship that may exist between them. Thus, the State continues to assume its fundamental role in the resources allocation to meet the needs of different sectors and regions.

In the long term, the future of the water issues will depend on the actual place of the irrigated agriculture sector in the general economy as the main consumer of available water resources. If the general trends observed today are confirmed, it is highly probable that, in 20–30 years, progress of industry and services will lead to a sharp decrease in the relative share of agriculture either in the general economy or in the working population. This situation, conditioned by the demand-resource balance, will lead to accept that the full mobilization of available resources can eventually modify the current conditions of access to water established between the different types of uses and between the different regions. Under these circumstances, more pertinent solutions should be found to the rural areas development, through improved rainfed agriculture and diversification of economic activities, as well as to the problem of food security by controlling food consumption and optimizing the country's trade balance.

It should also be noted that there is currently no systematic recognition of demand for environmental protection. After an unbridled mobilization and an accelerated exploitation of water resources, this deficiency will necessarily have to evolve in relation to the fragility of the natural environment and the likely impacts of climate change. As Tunisia develops, environmental protection is set to become one of the major concerns of water policies.

In the longer term, and in particular in the event of an imbalance between water supply and demand, more radical reforms of the water sector will have to be envisaged which will lead to a higher level of demand management, which seeks to increase the inter-sectoral efficiency of water use, with a focus on the most competitive sectors. In this context, "demand management in the strongest sense" is designed to disconnect water demand from economic growth and population growth. At an advanced stage, this disconnection uses instruments allowing for new reallocations of water resources, and such redistribution could favor the environment sector, sectors of high economic performance and those with a high strategic level, such as food security.

Like any major change, such a redistribution is likely to generate short-term social and political problems, owing to power relations, constituted lobbies and political influence, but it could still offer long-term orientations for reallocation of scarce water resources on more objective bases (competition and regulation between sectors, the extreme case of water markets) than for the allocation by force of administrative regulations where the State is the sole arbitrator.

This new step, which aims to put in place a policy favoring the search for increased water use efficiency, requires two prerequisites:

(A) The existence of a vision on the ways and possibilities of optimizing the use of water in all sectors. Such optimization is nowadays a tool for anticipating long-term challenges, by controlling the demand growth before developing new and increasingly costly resources, notably by integrating opportunities for regional development and determining the place of agricultural hydraulics in this development.

(B) The reinforcement of "adaptive capacity" for extreme hydrological situations and the capacity to manage situations of structural shortage, in which any additional resource mobilization, already almost fully exploited, becomes extremely costly. This capacity actually represents "all the social resources such as intellectual capital, scientific and technical skills, institutional capacity, financial capacity or the mobilization of individuals that a society can solicit to cope with the scarcity of water resources" (Treyer 2006). All these changes also require implementation of a strategy to bring about a "new water culture" that is up to the challenge. Among other objectives, this strategy should aim at a radical change of behavior, increased accountability of users and decision-makers.

5.2 The Drinking-Water Sector

5.2.1 Demand Management in the Drinking-Water Sector

As part of its demand management policy and to control water consumption, SONEDE has established over the past two decades an ambitious action plan with technical measures, financial measures, intensive media programs and public awareness campaigns for water savings.

The technical measures

The technical management of drinking water includes components relating to: (i) securing the water supply using the national water transfer system and the SONEDE regional interconnections networks; (ii) improvement of control systems in supply networks and tanks, and introduction of automatic management systems in extended networks; (iii) generalization of the counting system at the subscriber level; (vi) fight against network losses and leakage with improved materials, rehabilitation of dilapidated networks, pressure regulation in the appropriate areas and implementation of networks audit and leaks campaigns. These actions have had a positive impact on some of the networks technical performances. However, the overall network performance, which reached 81.4% in 2000, its highest index in the history of the national company, has declined since then, reaching 76.2% in 2010 and 70.7% in 2015. It is true that the produced water volumes have doubled over the same period.

About drinking water savings, they have long been perceived only through SONEDE direct actions on its networks or by the impact that its tariff system could have on the customers' behaviour. Individually, these measures are likely to prove insufficient if we really want to tackle water conservation, which is really at the uses level where a vast pool of water savings still needs to be explored and valued. We can evoke in this way the following actions:

(i) Technological innovation in the field of domestic water by imposing efficient standards and guidelines for local manufacturing and imported equipment.
(ii) With the collaboration of architects, widespread mandatory standards for civil, administrative and collective buildings and public institutions (schools, colleges, universities, etc.), which determine the most effective arrangements for water conservation: rainwater collection, possible recycling of treated waste water or gray water;
(iii) Introduction of water-saving aesthetic devices for watering public parks and private gardens (hotels, apartments, etc.), which are based on the use of efficient irrigation systems and plants species resistant to drought and with low water consumption;
(iv) Creation of a "center for water professions" specialized in all professional issues related to the water sector, necessary in view of the evolution of water issues in Tunisia.

The financial measures

The SONEDE pricing system for drinking water has evolved from a monomial system adopted in the 70s, to a tariff at progressive rates based on consumption brackets system, with a double objective: help socially disadvantaged categories to obtain water of satisfactory quality thanks to a reduced tariff, and financially penalize water overconsumption. The current system is of a progressive type depending on usage and consumption bracket with three categories: (i) domestic, public, commercial and industrial uses; (ii) tourism (hotels); (iii) public standpipes (Agricultural Development Groups supplying rural areas). The first category consists of seven quarterly

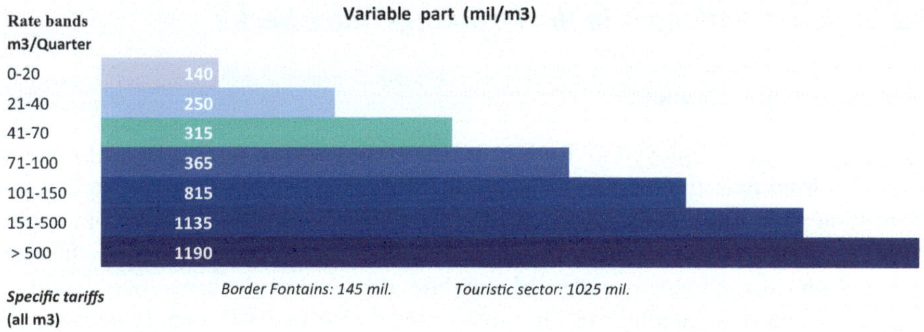

Fig. 5.1 SONEDE's pricing system for drinking water in 2014, 1 mil (millime) = 0.001DT (Dinar), 1 DT eq. US$0.55 on 31 Oct 2014

consumption tranches, each corresponding to a specific tariff. The fixed term of the binomial pricing is very low, fixed in relation to the diameter of the user's installed meters (Fig. 5.1).

In addition to the supply service remuneration, SONEDE users in areas covered by ONAS must pay the sanitation fee, which is also progressively priced and calculated on the basis of the quarterly water consumption. There are three categories of royalties according to use: Domestic, touristic and industrial. The quality of the discharge in the ONAS network is classified into three categories: low, medium and high pollution.

The binomial progressive pricing method is one of the key tools used by SONEDE as part of its demand management policy. Every consumer is, in fact, motivated by the pricing method to be careful not to exceed the upper limit of the usual consumption band. Exceeding the upper limit of a band leads to a substantial increase in the water bill, which is intended to encourage consumers to avoid waste and optimize and rationalize the use of water at their level.

Specific measures

Several specific measures have been initiated by the government to support water resource conservation programs and to encourage users to save water; the most important one is the adoption of the principle of periodic water audit, compulsory for large consumers, to evaluate the effectiveness of internal systems to regulate and rationalize the use of water.

These measures have introduced two major innovative orientations relative to the previous drinking water management policy: (i) give a capital interest to the users role in the resource preservation; (ii) the "breach" Inserted in the SONEDE monopoly on drinking water by allowing the private sector to participate in the satisfaction of public needs through non conventional water resources. The Water Code was amended by Act No. 2001-116 to integrate most of the necessary legal framework for the implementation of the above measures.

5.2.2 *Key Challenges in the Drinking-Water Sector*

Increasing water demand

Despite the means deployed in Tunisia to control population growth, it will remain relatively high over the coming decades and the population is set to increase from 10 million at the end of the years 2000 to 12.5 million in 2030. The rural population currently represents 35% of the total population, or 3.5 million, and should stabilize at 3 million over the next decades. Meanwhile, the level of the urban population is expected to increase again to reach a rate of over 75% in 2030, mainly because of people migration from inland areas to the coastal regions.

As an advantage for Tunisia, however, it should be noted that the present specific consumption of domestic drinking-water (90 L/day/capita.) is at a reasonably low level by international standards (Fig. 5.2). The WHO indicators advocate an average requirement for similar regions in North Africa of around 125 L/day/capita and 100 L/day/capita, respectively for individual connections and collective services. Apart from population growth, other factors such as general rise in living standards, a tendency to live in agglomerations, atomization of families by reducing the number of members and the increasing number of households, changes in individual behaviors, are likely to cause in the future a fairly strong growth of the specific consumption.

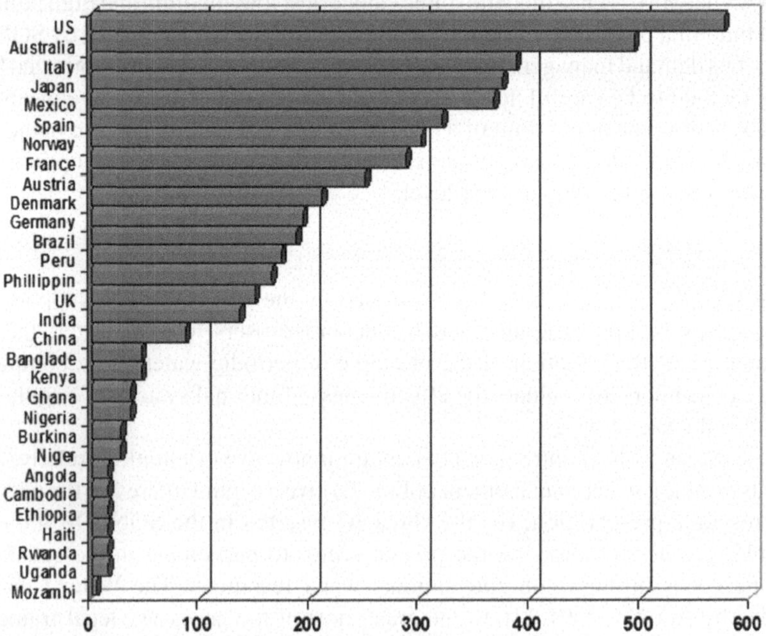

Fig. 5.2 Average specific consumption of drinking-water per country, in liters/day/capita. *Source* Data360 (2017)

The drinking-water salinity and standardization

Although the sanitary quality of the drinking-water distributed in Tunisia is relatively satisfactory, the high salinity level of water supplied in many regions raises gustatory and organoleptic quality problems often poorly accepted by consumers. It is estimated that nearly 7% of the population served by SONEDE receives drinking-water that, in some way, is not consistent with the provisional Tunisian standard NT 09.14. The latter is basically inspired by the guidelines of the World Health Organization in this field, with tolerances particularly as regards the maximum concentration of dissolved salts, fixed at a maximum of 2.5 g/L by the Tunisian standard although it is of the order of 0.5 g/L to WHO. The Tunisian standard was adopted as a transitional measure to allow drinking-water in areas where the quality, in terms of total concentrations, of the water resources is rare (the South region and some rural areas), and where it is not always possible to comply with the general norms. Although WHO normative guidelines are not mandatory, the draft Tunisian standard could not obtain the necessary internal consensus between stakeholders; it still retains, 30 years after its publication, a provisional status and has not yet been updated or objectively evaluated to assess its relevance.

Nevertheless, these "adaptive" provisions have allowed the drinking-water operators (SONEDE and Rural Engineering Authority) to ensure an honorable drinking-water supply to urban centers and rural areas. Overall, 98% of the population served by SONEDE receives water with a salinity of less than 2.5 g/L. However, this proportion drops to 93% if we consider the compliance with other quality parameters, including sulfates (less than 600 ppm content) and hardness (less than 100 °F). For information, the distribution of the total population served depending on the salinity of the water is as follows: 89% of the population is fed by waters with a salinity of less than 1.5 g/L; 6% get water with a salinity of between 1.5 and 2 g/L; 3% receive water of between 2 and 2.5 g/L; 2% are fed by water with a salinity greater than 2.5 g/L.

This has prompted SONEDE to revise its strategy and strengthen its brackish-water desalination programs for all areas supplying water with salinity above 1.5 g/L, especially in the regions of the South and South West. Such options may ultimately be favored in many other parts of the country. The systematic use of desalination to solve the problem of the saline drinking water would result in higher prices and would jeopardize the financial stability of SONEDE in the absence of a consistent pricing system.

Financial and institutional issues

(a) The dichotomous management of water services

The advent of SONEDE and ONAS, in 1968 and 1974 respectively, as public bodies and in a monopoly framework for the development of drinking water and sanitation on a large scale, have been a relatively successful bet for a large part of the national territory. This model of intervention has indeed been the basis for the extension of public water services in areas that are sometimes very remote and traditionally thirsty.

However, it is legitimate to ask whether the current dichotomous management model with separate drinking water and sanitation could keep its initial force for a long time more, and if SONEDE and ONAS would be able to maintain the same missions in the future, given the new nature of the constraints and pressures to which they are subjected.

In this context, a strategic reflection on water and sanitation in Tunisia (World Bank, 2009) resulted in recommendations for: (i) the creation of a national holding for water and sanitation; (ii) the establishment of a regulatory body for the coordination, guidance and evaluation of the sector; (iii) private sector participation (subcontracting, management delegation, public-private partnership). These findings deserve further study in order to initiate the institutional reforms necessary for the joint sustainability of the drinking water and sanitation sector.

(b) **The costs of drinking water and sanitation management**

Costs in the drinking water and sanitation sectors are constantly increasing and have often reached levels considered as critical. They will become worse in the future due to the need to undertake costly improvements to meet needs for infrastructures renewal and modernization and to reach the required level of water quality service (advanced techniques of treatment and desalination in particular). In Tunisia, tariff increases in drinking water and sanitation have a predominant social and political dimension and can not therefore follow the real costs increase in the production factors or in the services extension to marginal non profitable areas. Pricing systems for water utilities currently appear to have reached a limit for cost recovery and the financial balances of SONEDE and ONAS are already beginning seriously threatened without the government can provide financial compensation inherent to the official policy. This situation, unsustainable in the medium term for a public company, risks putting the current Tunisian centralized public drinking water management model into great difficulty.

A number of reforms in the drinking water sector deserve to be initiated in the medium and long term and should concern:

(i) Introduction of the water resources as a basic factor element for spatial planning and regional development. Water has become a scarce and expensive commodity and land-use planning will need to transparently reflect the availability and value of water resources at the local level, as well as the cost-effectiveness of the investments to be made in case of possible water transfer.
(ii) Rethinking SONEDE's current drinking water pricing model and the equalization conditions between consumption sectors so that they better take into account, on the one hand, strictly domestic consumption with its social component, and on the other hand other consumptions of economic and competitive nature. The latter, relating to industrial, touristic and commercial activities, should support the real costs and priced according to the economic value of water. It should be noted that regional differentiation of tariff systems is currently a reality in rural areas managed by user's associations. Each drinking

water GDA is required to set its own water tariffs based on the local level cost of managing its water system.

(c) **The duality of the drinking water supply system**

SONEDE operates in agglomerated areas, urban or rural, and drinking water GDAs (1250 in 2014) operate in scattered rural areas (15% of the total population). Due to urbanization development, SONEDE will have to integrate in the future an increasingly large part of the rural areas currently under the competence of the GDAs, which will involve major investments for the conversion of the networks, then an increase of the water demand due to extension of individual connections and the consequent need for sanitation. These developments will be difficult to manage for SONEDE without substantial financial support from the State. Only drinking water systems in rural areas with dispersed populations managed by GDAs may have some favourable conditions to sustainability on the long term, as they retain a multiple role: human food supply, livestock supply, complementary irrigation and drought control, which are considered essential for agricultural and rural development in arid and semi-arid areas.

(d) **The rural sanitation issue**

In rural areas, the development of sanitation is significantly delayed with respect to the urban environment, with serious risks to populations' health and possible soil and water contamination at the territorial scale. This situation is due, on the one hand, to the lack of a clear institutional framework for rural sanitation and, on the other hand, to the absence of a concerted national policy in this area, defining the types of technical interventions and financial resources. Due to the profound improvement of rural drinking water supply, it is desirable that a joint improvement can be achieved in terms of sanitation in a realistic timeframe.

5.3 The Irrigation Sector

5.3.1 General Configuration of the Irrigation Sector

Irrigation has always played a vital role in the agricultural and economic history of many regions of Tunisia, and the links between water and agriculture now occupy a prominent place: (i) the agricultural sector accounts for a significant share of water resources and irrigation accounts for almost 80% of the total water demand; (ii) a significant portion of the increase and diversification of agricultural production will result from irrigated agriculture, seen as a key challenge in food security; (iii) irrigation is one of the privileged elements of rural development and land-use planning and a lever for balancing the agricultural economy particularly in drought periods.

Significant public and private investments have therefore been made to exploit the water resources and to develop irrigation schemes. Within the limits of water and soil resources usable for irrigation, potential irrigable area with full or partial control irrigation equipment is of the order of 496,000 ha, or 8% of the agricultural area. The potential for semi-intensive irrigation, including flood spreading, is estimated in a very comprehensive way to 150,000 ha. The actually irrigated physical area is only 407,000 ha. The extent of irrigable surface areas has been on the order of 1.9% over the last decade.

Intensive irrigated public schemes, with collective infrastructure provided by the State, cover an area of 254,000 ha (51% of the irrigable total area) and are fed mainly from dams, deep wells and recycled wastewater (7000 ha). Private irrigation schemes, carried out by farmers, amount to 242,000 ha (49% of total irrigable area) and are mainly supplied from dug wells, deep private boreholes and facilities to catch flash floods on wadis. Overall, the role of the State in the development of private schemes remains paramount; thanks to a system of financial investment incentives for irrigation management on farms, as well as valuing irrigated land in general.

Irrigable areas are expected to stabilize over the next years and newly developed areas should only compensate for the losses caused by soil salinization in some areas due to the use of poor quality waters on the one hand, and by the local reduction of water withdrawals as a consequence of the intensive overexploitation of a certain number of aquifers.

The annual water demand of the irrigated sector depends on many factors but it is particularly sensitive to the increase in irrigated area and to the climatic conditions of the current year. The demand increased from 1575 million m^3 in 1990 to 2210 million m^3 in 2010. The average water demand per intensively irrigated hectare is estimated at 5500 m^3, but may go from 1000 to 2000 m^3/ha for cereals and fodder in the North to reach 15,000 to 20,000 m^3/ha for date palms in the southern oases.

From an economic point of view, with 8% of agricultural land, irrigated agriculture now accounts for 35% of the total value of agricultural production, for 90% of vegetable production, 75% of fruit production and 30% of dairy products. It contributes to 20% of the value of agricultural exports and accounts for 27% of agricultural employment.

In the early 90s, the irrigation sector adopted a water-management policy of rationalization and search for greater efficiency; to establish the first measures for managing water demand, guidance became essential in order to maintain the demand for irrigation water at a level compatible with the available resources and prepare the sector for a competition which might be to its disadvantage.

5.3.2 The Instruments of Water Demand Management

The irrigation sector's demand management strategy is based on a number of methods and instruments: (i) transition from a technical management approach to an integrated one; (ii) rehabilitation and modernization of collective networks to improve the efficiency and quality of service; (iii) establishment of a participatory approach with increased accountability of users when creating new irrigated areas and during operating and maintenance phases; (iv) implementation of a water pricing system for a progressive upgrading of cost recovery and combining flexibility with food security objectives; (v) financial incentives for the promotion of modern and efficient irrigation techniques on the farms.

Modernization of collective irrigation systems and improving efficiencies

With a contemporary design and mainly set up in the 70s and 80s, public irrigation systems have gradually become obsolete and degraded, often through lack of maintenance and repair of irrigation and drainage equipment and associated facilities (paths, windbreaks), with adverse consequences for the quality of service to irrigators and thus, reducing the level of agricultural intensification. A major effort has been made over the last two decades to initiate comprehensive programs of rehabilitation and modernization of the most important systems.

As the operations of rehabilitation/modernization have become cyclical, they are fully supported by the Administration. To minimize the cost, institutional and financial solutions will always be found to ensure continuous and preventive measures deemed less costly and to involve user fees generated by this necessary modernization. For the whole country, one can estimate at 60% the present efficiency of old public systems of gravity sewers, being modernized, and 90% that of recent-design pressure systems. An overall average efficiency of about 80% can be considered realistic for all public networks in the country. The long-term objective is to achieve a rate of 95% after a comprehensive modernization and application of sound management of water systems.

Water savings at the farm level

While significant efforts have been made since the 70s to promote modern irrigation techniques directly to the plot, the traditional gravity irrigation has persisted in many irrigation schemes, even of recent design, causing huge water losses that severely limit the agricultural intensification efforts of developed land. A new approach was initiated in 1993, consisting in tackling the problem in an integrated manner: extensive outreach programs for farmers, extension-specific training for technicians and engineers of the irrigation sector, the search for revitalization-development for efficient irrigation techniques, establishment of an adequate organizational framework at national (national monitoring and evaluation unit) and regional level (intervention unit and local coordination), the private sector involvement (equipment and service suppliers).

The National Water Saving Programme (NWSP or PNEE) aimed to encourage all irrigating farmers to reduce water wastage at the farm level. This programme has experienced considerable momentum from 1995, favored by the increase in investment bonuses for efficient irrigation equipment: from 30%, these bonuses were raised to 40, 50 and 60% respectively for large, medium and small farms. In terms of water-savings equipment in irrigated areas, the results were encouraging. Nationally, 85% of the total irrigated area was equipped in 2014, three-quarters of which by the most modern techniques of water saving: sprinkler and drip irrigation (Fig. 5.3).

A special feature of this program was not to focus on one single irrigation technique of localized irrigation potentially considered the most efficient, but to open the way to other techniques, for example, spraying or enhanced gravity track (canal lining using pipes, building control tanks, etc.) as a function of local conditions (nature of the crop, water quality, economic efficiency), and of the capacities of irrigators to monitor modern techniques. This diversification may explain the size reached by the program of large irrigated areas.

The overall assessment of the NWSP carried out in 2016 shows, inter alia, that half of the farms have stabilized their water consumption, one third have declined, and a small minority have increased their water withdrawals. These variable results in terms of reduced water use are to be compared with a sharp increase in crop yields in 70% of farms. Similarly, the economic valuation of irrigation water has to the minimum doubled for all speculations. The NWSP has therefore led to a significant increase in water productivity, which more than doubled in 20 years for fodder, market gardening and tree farming, and particularly for green barley, chilli, vines and apple tree.

Criticism of the NWSP is generally twofold: (i) the high level of financial incentives given to modern irrigation equipment; (ii) second, the use of saved water volumes to extend the irrigated areas. As far as financial incentives are concerned, they were reduced from 2007 to favor only farms to be equipped for the first time. As for the extension of irrigated areas by saved water volumes, this extension is in fact sought to improve land intensification in public irrigated areas where water allocations are physically fixed and controlled. When the source is groundwater and in case of aquifer overexploitation, this extension is certainly not sought and is considered as a perverse effect of the program.

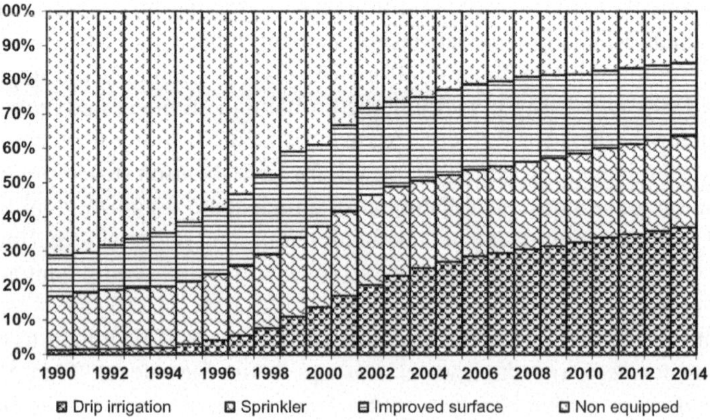

Fig. 5.3 Share of different areas equipped with modernized irrigation between 1995 and 2014 (DGGREE 2016)

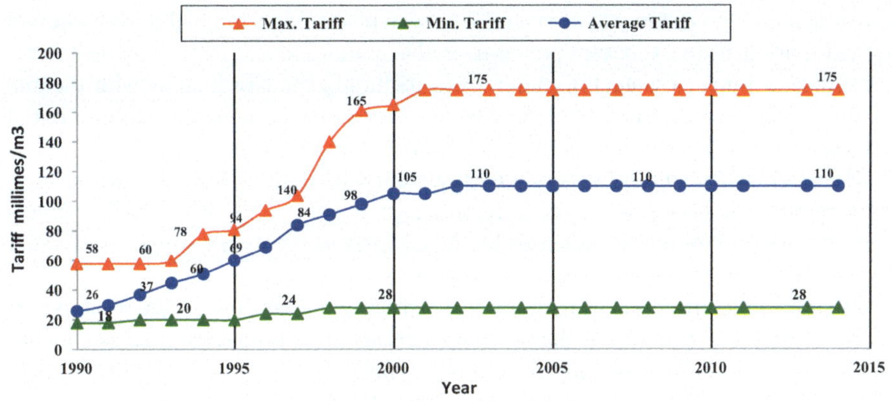

Fig. 5.4 Change in the pricing of water irrigation for the period 1995–2014: maximum charge, minimum charge and average charge (DGGREE 2015)

Besides the development of modern equipment to reduce losses and tackle the technological backlogs of irrigated farms, other options exist that would significantly improve the efficiency of irrigation water use; for example:

(i) Improving irrigation management conditions, combining information about the local climate, soil and the crop, in order to arrive at a precise irrigation;
(ii) The adoption of deficit irrigation, applying water at predetermined stages of the development or growth of the plant to save water while improving the efficiency and quality of the production (Ben Nouna et al. 2005).

In reality, the improvement of water management at the scale of the farm is still in its infancy and the possibilities of reducing consumption and a utilization of scarce resources constitute the real challenges for the future of the irrigation sector.

Reforms of the pricing policy

Reforms of irrigation-water pricing were initiated by the Offices for Irrigated Perimeters in the 80s, with the triple objectives of price and cost transparency, flexibility (regionalized pricing, changes in rates according to the vocation of irrigated areas) and related national socio-economic objectives (revenue enhancement, food safety, etc.). But it was by the Agricultural Structural Adjustment Program of 1986 (PASA) that tariff reforms were implemented most vigorously. From 1990 to 2000, a steady increase in water tariffs by an average of 9% a year in real terms was adopted (Fig. 5.4). Alongside these measures, considerable efforts were made to generalize the volumetric pricing of water, by introducing metering systems on irrigated farms.

These increases have served to absorb a significant portion of the higher operating and maintenance costs of water systems, especially in large irrigated areas. The recovery rate of these expenses rose as well, for the same period, from 57 to 90%. Aware of the limits of the monomial tariff (fee proportional to the volume of water consumed), the government has since 1999 been proposing a gradual introduction of the two-part tariff (a fixed term per cultivated hectares and a term proportional to the consumed volume of water) in major areas including the North areas with the aim of improving the recovery rate of the cost of water and encourage the exploitation of irrigated land still partly remaining within the rain-fed system.

However, as a result of farmers' disagreement with their GDAs, this tariff policy, considered to be very restrictive, suffered a significant decline from 2002 onwards with the stopping of tariffs increases by the CRDAs as wholesale suppliers of water Irrigation in public irrigated areas (see Fig. 5.4).

Despite some progress, agricultural water pricing remains one of the weak points of the current water policy, as the tariff systems are sometimes designed as a means of allocating an indirect subsidy or improving farmers' incomes, while they should also be a powerful policy instrument for irrigated agriculture policy.

The state has invested substantially in the development of irrigation schemes without any effective participation of the beneficiaries, and continues with GDA to provide water at prices that do not reflect its economic value in a water scarce country. This provides no incentive to preserve the resource and only conveys a message contradicting the one that advocates the installation of equipment to save water at the plot. Agriculture certainly needs to be supported to improve the income of producers, but this can be planned in other, more favorable contexts and without affecting the viability of the sector. Also, is it important to establish a fair and efficient pricing policy to ensure the autonomy of local management and the means needed to maintain the structures and the hydraulic equipment necessary for the sustainability of the productive potential of irrigated areas.

Participatory irrigation management

Since 1987, several components related to agricultural development have been decentralized and regional or local structures to oversee and stimulate this development have been strengthened or introduced: CRDA regionally, GDA locally. Participatory management is a key concept in the approaches to improve the efficiency of water management in conditions of scarcity. "Participatory irrigation management" means that the farmer—the user—participates in all aspects and at all levels of irrigation management.

The experience, started in Tunisia, of total participatory management, resulted in a transfer of responsibility and authority to user organizations for the governance, management and financing of public irrigation schemes. This transfer has required a number of prerequisites and accompanying measures:

(i) The political decision to transfer a significant amount of responsibility for irrigated perimeter management for the benefit of water users, through the structural adjustment program in 1986;
(ii) A legal framework for the establishment of independent associations with real options;
(iii) The development of technical, administrative and financial capacities in associations, able to manage irrigation systems;

(iv) The ability of public administrations or agencies responsible for irrigation sector associations to provide the necessary institutional and technical support and supervise their performance (specialized services or units within the DGGREE and CRDA);
(v) An economically viable irrigated agriculture authorizing the financial independence of associations, an economic viability which should contribute to all forms of support and incentives for investment in irrigated agriculture.

In 1992 participatory management experienced a new impetus with the adoption by the State of a specific national strategy for the creation and monitoring of associations, to promote water-system management autonomy as part of regional irrigation schemes and rural drinking-water supply. This option was dictated by the policy of structural adjustment in the agricultural sector, causing the State to withdraw from many functions. While keeping the preeminent role of the State in the investment in water infrastructure, the strategy should ensure:

(i) a gradual disengagement of the Administration from direct management of public irrigated areas, transferring exploitation and infrastructure maintenance to Agricultural Development Groups (GDA);
(ii) reduction of direct and indirect subsidies to bring these organizations, in the long term, to pay the real price for water: operation and maintenance costs;
(iii) Technical support to GDA allowing them to participate more dynamically in managing infrastructure and developing the agricultural perimeters.

These groups numbered around 200 in 1995 and 1000 in 2000. In 2014, there were 1248 GDA who managed, to varying degrees, almost all of the public irrigated areas, of which 88% are really active; the remaining 12% have temporarily ceased their activities for technical or social reasons. These GDA exploit 60% of the total water volume allocated to irrigation.

The management of water infrastructure by membership organizations enabled users to realize the potential benefits of this new mode of self-management, including ownership in the longer term, of these infrastructures. The, quite relative, observed successes are the result of a wide-spread awareness and an almost continuous assistance provided by the CRDA to GDA: awareness campaigns among members, technical, administrative and financial support. In 2012, it was estimated that almost 70% of GDAs were able to a more or less autonomously meet the necessary expenses for operation and maintenance of water systems through fees. These conditions have been achieved in many medium and small farms, but remain difficult to achieve in larger facilities.

A number of GDA encounter serious social, technical or financial difficulties, which limit their autonomy and action efficiency. These challenges: conflicts between users, inability to pay for outside services, technical malfunctions of water systems, have generated strong opposition to the associative movement and the complete cessation of activity of certain associations. It is true also that the pressure from local and regional administrative and political authorities has long maintained suspicion and mistrust among the members of GDA and prevented the emergence of a democratic form of governance within the organizations.

During the 2011 revolution, the GDA were subject to unprecedented social pressures and experienced a serious crisis of confidence. The vast majority of executive and management bodies in place were dissolved and replaced by interim committees. The most worrying dysfunction indices are today the derisory level of fee collection and its corollary, indebtedness has become excessive and unsustainable for water (SONEDE) and energy (STEG) suppliers. To these factors is added a level of water pricing clearly insufficient for a sound operation of the water systems and consequently recourse to the services of CRDA for all maintenance and operations.

5.3.3 The Irrigation Constraints and Main Challenges

Despite improvements in the performance of the irrigated sector following the various economic and institutional reforms undertaken during the last two decades, several constraints of a structural or cyclical nature still hinder its development in an optimal framework, and several challenges risk undermining its favorable development. Even if it is favored by the more or less continuous availability of the water factor, irrigation remains an environmentally fragile and highly capital-intensive activity at the economic level. Its development is intimately linked to the general context of agriculture and the agrifood sector, domestic agricultural markets and exports. We present below the most significant questions.

Viability and sustainability of productive assets

In a fragile context, a number of constraints related to irrigation risk affecting the sustainability of productive assets, namely: (i) groundwater overexploitation, (ii) salinization and degradation of irrigated soils.

(a) **Overexploitation of groundwater resources**

Groundwater overexploitation is affecting nearly a quarter of the irrigated perimeters of the country and in particular the private perimeters. It constitutes an essential risk factor for marine intrusion in eastern coastal aquifers. As for irrigation with weakly renewable groundwater, it is predominant in the South oases, where irrigation uses about 550 million m^3, that is to say a quarter of the total water supply to irrigation. Signs of degradation (excessive drawdowns, loss of artesianism, water quality degradation) are observed in a worrying way for the future of certain cultivated areas.

Several water-saving programs are being implemented to reduce water wastage in private areas and oases, but the future will remain uncertain as long as an integrated approach to the economic and social development of these areas, particularly southern oases, will not be implemented to reduce the pressure on the available scarce water resources. A groundwater participatory management is already beginning to move, although very slowly.

(b) **Salinity and degradation of irrigated soils**

Water used for irrigation in Tunisia is generally of medium to high salty quality, which affects the level of crop yields, but may also degrade, in the long term and sometimes in an irreversible way, the structure and fertility of irrigated soils. This phenomenon is often exacerbated by excess water or waterlogging caused by over-irrigation and an exaggerated rise of the water table.

Research on the impact of irrigation water salinity has been well developed in the 60s in Tunisia: the Center for Research on Use of Salt Water in Irrigation (CRUESI 1970) and its successor INRGREF accumulated practical experience in the different regions on the use of salt water in agriculture and accompanied the development of several irrigated public perimeters.

Of the total area currently equipped for irrigation, there are 60,000 ha of which the soils are sensitive to salinization in all its forms (primary and secondary) and 75,000 ha are subject to water table rise. In terms of perimeters, it is estimated that 60% of irrigated public perimeters are moderately sensitive to secondary salinization due to irrigation; this rate reaches 86% in private perimeters (MARH-DGACTA 2005).

Presently, the tendency to use, for irrigation, saline natural waters or increasingly saline drainage waters carries many risks, while the control of lands already managed sometimes poses serious problems in terms of sustainability: inadequate cultural practices to high salinity, insufficient leaching volumes, inadequate drainage systems, etc.

Agricultural valuation of available water resources

Current levels of irrigation water use remain relatively low in relation with efforts to improve irrigation schemes. Levels of agricultural intensification, both horizontal and vertical, often remain insufficient, and the macroeconomic performance of irrigation is still perfectible. The irrigation sector still holds real possibilities not yet expressed in many regions. Horizontal intensification in some public irrigation schemes is still modest, and the crop rotation recommended for land use is often poorly respected.

An analysis of the achievements in terms of actual irrigated area shows that the average rate of crop intensification (ratio of the total area actually irrigated during the crop year to the total irrigable area) is about 90% but varies between 40 and 120% depending on the perimeter. On the other hand, the levels of yields achieved in irrigation are relatively small and very variable from a perimeter to another and within the same perimeter. We can estimate that current yields would be about 50% in relation to potential irrigation yields.

The improvement of these performances will depend on the ability to remove the endogenous and exogenous constraints that still weigh heavily on the future of the sector. Indeed, irrigated agriculture is inserted in a socio-economic context that affects its profitability. The main determinants of this profitability are of several types: (i) the proper functioning of agro-food chains and agricultural markets as well as the external trade regimes; (ii) the availability of water in a sustainable manner at the local level, the cost of water given the changes in the conditions of exploitation of the resource and the evolution of the cost of energy, etc.; (iii) the professional structuring of farmers and their ability to master technologies related to water management. These contextual elements, which are inherently evolutionary, remain based on the strategic and development challenges of Tunisian agriculture as a whole.

5.4 Reuse of Treated Wastewater

5.4.1 Context and Constraints of Treated Wastewater Recycling

In water scarce regions, reuse of treated wastewater, or Recycled Water (RW) is generally considered both to compensate for water shortages and ensure a better water quality in receiving environments. As a water supply project, RW is subject to a number of constraints (AFD-BRLi 2011), the most important being:

- Water demand for agriculture is cyclical, whereas RW supply is virtually regular throughout the year, which means considering regulation systems, in particular inter-seasonal storage.
- In agriculture, the microbiological quality of RW can constitute risks of direct contamination for irrigators and raw products consumers.
- The chemical and mainly organic quality is an advantage with the presence of agricultural nutrients, but can become a source of nuisance to the soil by the salinization that can sometimes accompany RW.
- RW must be made acceptable with regard to the health risk; for that purpose, precise scientific knowledge of the effects of various treatment processes and all stages of the hydro-biological and anthropogenic RW cycle is required.

In Tunisia, the acquired experience in RW dates back to the beginning of the sixties with the development for the Soukra irrigation area near Tunis on 1200 ha. Experience has consolidated over time with the increase of treatment stations managed by ONAS. By 2015, the treated wastewater production is 242 million m^3 and the recycled volume 68 million m^3, i.e. a recycling rate of 28%. In 2008, the RW volume totals 57 million m^3 distributed according to the following uses (AHT Group 2009):

1. 29 Public perimeters irrigated with RW (DGGREE 2008), on a 5500 ha area, of which 3000 ha are effectively irrigated and use a volume of 22 million m^3/year.
2. Irrigation of 10 golf courses with a total area of 1300 ha, irrigated with RW, using 10 million m^3/year.
3. Watering green spaces: consumes 7 million m^3/year; includes in particular a number of hotel units and the green areas of the main touristic roads; totals 450 ha.
4. Ecological reuse in wetlands: include sebkha Kelbia and the sebkha of Korba; totals 2 million m^3/year.
5. Aquifer recharge: Oued Souhil and Korba aquifer; totals 0.62 million m^3/year.

Table 5.1 Treated wastewater recycling in 2008, according to ONAS (2010) and AHT Group (2009)

Area of use	Quantities
1. Irrigated perimeters	22 Mm^3/year
Area equipped	5500 ha
Effectively irrigated area	3000 ha
2. Golf courses	10 Mm^3/year
Watered area	1000 ha
3. Green spaces	7 Mm^3/year
Watered area	450 ha
4. Wetlands	2 Mm^3/year
5. Recharging aquifers	0.62 Mm^3/year
6. Irrigation by Medjerda	15 Mm^3/year
Total Treated wastewater reuse	57 Mm^3/year

All these uses represent a total volume of 42 million m^3/year. Table 5.1 summarizes all RW uses. The complement to 57 million m^3 (15 million m^3/year), represents the indirect water withdrawals after mixing of treated wastewater in Oued Medjerda, downstream large cities in the basin (ONAS 2010).

5.4.2 *Main Challenges*

Irrigation with RW continues to be problematic, as the treated water quality, although essentially of domestic origin, remains highly variable. This situation also limits uses of these waters by other potential sectors (municipal, industrial, touristic, etc.). This recycling, long encouraged to protect the touristic coastline (82% of effluents are discharged at sea), is still below expectations despite the introduction of a modern irrigation infrastructure. A fairly comprehensive regulatory framework based on the WHO and FAO guidelines on the conditions of agricultural use of RW, as well as tariff incentives (single incentive tariff of 0.02 DT/m^3, i.e. 20% of the average tariff of conventional agricultural water).

As part of its policy of enhancing the value of RW, ONAS has already undertaken a study on the cost-effectiveness of the different reuse options for treated wastewater in sectors other than agriculture. The analysis showed that these sectors still have reservations about the reliability of supply and the quality of water. ONAS has even initiated a study on the conditions of long-distance transport of treated wastewater from Greater Tunis (producer of about 50% of treated wastewater nationally), for irrigation and groundwater recharge, to the agricultural regions of Zaghouan and Kairouan.

Despite the experience and the general framework conducive to recycling, a number of constraints continue to hinder its large-scale development. These constraints are numerous and include:

(i) The physical and chemical quality of treated wastewater, often considered unsuitable for reuse, because treatments are sometimes inadequate, in particular against high salinity, resulting from the quality of the original water;
(ii) The variable, and not always adequate, wastewater treatment may have an effect on human health. ONAS has long since adopted a level of general secondary wastewater treatment that only allows restrictive uses for agricultural purposes;
(iii) Maintaining an acceptable level of intensification in areas already equipped for irrigation with the treated wastewater requires an adequate investment policy for the modernization, rehabilitation and renewal of old treatment plants;
(iv) Irrigation still represents the majority of RW but irrigation is not able to recycle all the possible production of treated wastewater in the long term.

Presently and for the medium and long term, ONAS seems to decide for the consolidation of acquired knowledge by improvement of the secondary treatment and setting up of some tertiary treatment pilot units.

5.4.3 Opportunities for the Development of Recycled Water

Tunisia is regarded as a relatively advanced country as to treated wastewater recycling but successful international experiences help enrich the Tunisian experience. The reuse of treated wastewater has reached high levels in arid countries, for instance Jordan and Israel mainly for agricultural purpose. The Californian experience is instructive as regards the diversification of uses where only 46% of treated wastewater is recycled for irrigation, 21% for recreation and leisure (landscape, recreational lakes, etc.), 14% for aquifer recharge and 19% for various other uses.

Among treated wastewater uses that may be tested on a large scale and then developed, one may cite the following possibilities: (i) Adequate mix of conventional and treated wastewater for unrestricted water use in agriculture or in some industrial sectors; (ii) Direct use of gray domestic water from washing machines and kitchens for watering private gardens; (iii) Special treatments of wastewater for municipal needs of cities (landscape irrigation, fire-fighting, street cleaning, feeding recreational lakes, etc.); (iv) use in aquaculture; (v) integral treatment of wastewater for the production of drinking water (Schroeder et al. 2012).

The development of new uses of treated wastewater and their wide dissemination require R&D programs. Appropriate actions are required to know how to adapt the different technologies, evaluate their economic feasibility and establish the basis of regulations and standardizations. One should also observe that recycling treated wastewater with all its varied and complex operational issues, currently presents some difficulties related to promoting, monitoring, assessment and awareness that cannot be the sole responsibility of ONAS, whose action in the field of water purification remains essential. This institutional issue should be clarified to achieve a coherent and effective policy for treated wastewater recycling. Moreover, Tunisia is a touristic country and an exporter of agricultural products, a situation that should argue for a well-controlled recycling of treated wastewater, avoiding all major risks.

5.5 Water Desalination

SONEDE (2004) defined an optimal scheme for water supply to Cape Bon, the Sahel and the Southeast which prepared a long-term outline of the national desalination program. The best among the proposed solutions were those based on large-scale seawater desalination. According to this strategy, SONEDE abandons additional water transfer programs from the northern system and plans to build large seawater desalination plants in Sfax, Gabes and Djerba.

5.5.1 Achievements and Projects

In 2011, the total desalination capacity installed by SONEDE was of the order of 72,000 m^3/day, and the volume of produced desalinated water was 19.3 million m^3. Furthermore, private desalination plants (hotels, industries) have an additional overall capacity estimated at nearly 50,000 m^3/day, (Zaara 2008). As part of the drinking-water saving strategy, financial incentives were decided in 2001 in favor of the private sector with the aim of enhancing local water desalination. All of these programs should combine to reach a total production of desalinated water of around 40 million m^3 in 2010 which represents nearly 2% of volumes desalted in the Mediterranean Basin, (Blue Plan 2010).

Various programs of brackish and seawater desalination are being launched or announced for the future, (Nouicer 2012): (i) The national program to improve the drinking-water quality by desalination of brackish water in the South and the Center (Tozeur, Kebili, Gabes, Medenine, Gafsa, Sidi Bouzid): phase-I (10 stations, capacity 36,200 m^3/day) and phase II (8 stations, capacity 32,500 m^3/day); (ii) The planned concession project (building and operating) of seawater desalination in Jerba, (capacity 50,000 m^3/day), (iii) The concession (Gabes and Medenine), in Zarrat desalination plant, (capacity 50,000 m^3/day), (iv) Sfax seawater desalination project (capacity 150,000 m^3/day).

Furthermore, desalination units combined with renewable energy sources have been built or planned, most of which are experimental and intended for isolated communities in different regions (Beni Khiar: 3 m^3/day, Ksar Ghilène 15 m^3/day and Ben Guerdane: 1800 m^3/day). To meet the requirements of very good water quality, two desalination plants in connection with high added-value agricultural systems were built, one for greenhouse irrigation with geothermal water in Gabes for early vegetables production, the second in Kairouan for the production of aromatic herbs for export. Although the Tunisian experience in desalination is still limited, one can make the following remarks: (i) the use of reverse osmosis membrane technology is almost universal; (ii) to date, large-scale desalination concerns exclusively brackish water in the South; the progression to seawater desalination is envisaged in the short term with the next major coastal projects in the Southeast; (iii) major project-financing constraints seem to steer the government towards public-private partnerships for the production of desalinated seawater; management and distribution of desalinated water is still the responsibility of SONEDE.

5.5.2 Desalination Development Prospects in Tunisia

It must be noted that the Water Sector Study (DGRE 1999) did not make desalination a firm option for alternative resource development. The prospective study EAU 21 (Louati et al. 1998) predicted the production to be only 49 million m^3 in 2030. As for the strategic study Water 2050 (ITES 2011a), it has developed a vision of the volumes to be desalinated that should cover the shortfall once all water resources have been inventoried: the required volumes of desalinated water would then be 315 million m^3 in 2030 and 480 million m^3 in 2050. In the absence of a clear national vision of the role to attribute to desalination in the development of additional freshwater resources, it is the strategy of SONEDE that has the merit of giving an overview of future desalination programs in Tunisia. Starting from the already completed programs and extrapolating the drinking-water needs, SONEDE estimates that producing water by desalination will reach 117 million m^3 in 2030, almost three more time the 2010 production. Although this volume represents less than 3% of usable fresh water, the desalination of brackish water and seawater is now presented as one of the pillars of the national strategy and should represent nearly 20% of the drinking-water resources by 2030. But public water policies have not counted on such a rapid change and the country is not sufficiently equipped to confront a challenge of this magnitude.

Need for a specific desalination policy

The country does not seem equipped to face all the challenges that arise from such a perspective, but a number of guidelines and measures taken in a timely manner, may help better to control these developments; these measures include the following: (i) designing appropriate regulations and a standardization for the installation of desalination facilities, especially concerning environmental aspects, (ii) setting up an institutional organization able to plan, implement, evaluate and monitor the policy of public organizations and private sector interventions in the field of desalination, and define the mechanisms for technical assistance and financial incentives to accompany the policy.

References

AFD-BRLi (2011) Réutilisation des eaux usées traitées; Perspectives opérationnelles et recommandations pour l'action, Rapport Final; Agence Française de Développement and BRL ingénierie

AHT Group (2009) Identification et élimination des goulets d'étranglement pour l'utilisation des eaux usées dans le cadre de l'irrigation ou autres usages, RG/2008-01/FTF; Rapport National Tunisie; EUROMED; BEI; MEDA

Ben Nouna B, Zairi A, Ruelle P, Slatni A, Yacoubi S et al (2005) Evaluation de la demande en eau et pilotage de l'irrigation déficitaire des cultures annuelles: méthodologie et outils de mesure. Séminaire sur la modernisation de l'agriculture irriguée, 2004, Rabat, Maroc. IAV Hassan II, 15 p, 2005. <cirad-00189106>

Blue Plan (2010) Le dessalement de l'eau de mer: une réponse aux besoins d'eau douce en Méditerranée? Les Notes du Plan Bleu, Environnement et Développement en Méditerranée, N° 16, Juillet 2010

CRUESI (1970) Recherche et formation en matière d'irrigation avec des eaux salées; Rapport technique 1962–1969. Tech. Rapt/Unesco/Undp (SF) Tun 5

Data360 (2017) http://www.data360.org/dsg.aspx?Data_Set_Group_Id=757. Accessed 31 Oct 2017

DGGREE (2008) Situation des périmètres irrigués à partir des EUT. Décembre 2008

DGGREE (2015) Ministère de l'Agriculture, des ressources Hydrauliques et de la Pêche. Evaluation des activités des Groupements de Développement Agricole relatifs à l'irrigation

DGGREE (2016) Rapports annuels du système de suivi de l'exploitation des périmètres publics irrigués

DGRE (1999). Etude du secteur de l'eau. Groupement Bechtel./Scet -Tunisie

ITES (2011a) Etude stratégique - Eau 2050, Annexe 2, Mobilisation des ressources en eau

Louati MH, Khanfir R, Alouini A, El Euchi ML, Marzouk A, Frigui HL (1998) Eau21; Stratégie du secteur de l'eau à long terme; Ministère de l'agriculture, Tunis

MARH-DGACTA (2005) Examen et évaluation de la situation actuelle de la salinisation des sols et préparation d'un plan d'action et de lutte dans les périmètres irrigués de Tunisie, Programme PISEAU

Nouicer A (2012) Le dessalement en Tunisie, Sharing knowledge foundation, Gammarth, 17–20 may 2012, https://www.sharing-knowledge.org/wp-content/uploads/2017/07/Pdf-Nouicer.pdf

ONAS (2010) Rapport d'activité 2010

Schroeder E, Tchobanoglous G, Leverenz HL, Asano T (2012) Direct potable reuse; benefits for public water supplies, agriculture, the environment, and energy conservation, NWRI White Paper

SONEDE (2004) AEP du Cap Bon, Sahel, Sfax, Gabès, Médenine et Tataouine, Etude de faisabilité –Rapport définitif de mission 2

Treyer S (2006) A quelle raréfaction de l'eau faut-il se préparer? Construire une intervention prospective au service de la planification pour les ressources en eau en Tunisie, Thèse de doctorat de l'ENGREF Paris

World Bank (2009) République Tunisienne, Réflexion stratégique sur l'eau potable et l'assainissement en Tunisie - BIRD/Rapport 44744-TN

Zaara M (2008) Le dessalement en Tunisie pour l'amélioration de la qualité de l'eau potable desservie; RMSU MEDA Water Programme, ADIRA Workshop, Marrakech

Chapter 6
Water Security in Tunisia: Debated Issues

Contents

After having dealt with water policies and examined issues relating to the various resources and their different uses, we propose in this chapter to address some of the water issues still being debated, issues for which no solution has yet been envisaged with any determination or for which tentative solutions have been mooted but have not yet obtained a consensus among water experts, water actors and the civil society. The questions that preoccupy Tunisia serve to organize the debate around the main two following themes:

(a) **How to secure the water supply?** This theme covers physical and technical issues that the Tunisians, engineers and decision-makers traditionally know how to formulate and resolve. All the objectives that this theme includes do not, in fact, represent major controversies and might even be consensual; the views expressed reflect divergences concerning the opportunities, timelines, methods of approach and implementation;

(b) **How to achieve good water governance?** This theme covers legislative, institutional, regulatory and cognitive issues, most of which are subject to wide-ranging

M. Besbes et al., *National Water Security*,
https://doi.org/10.1007/978-3-319-75499-4_6

debates initiated some twenty years ago and formalized by the Tunisian water community. Most of these issues are still open to debate. It is true that energetic action has been taken to boost the water policy, in particular by strengthening water demand management in the various sectors through the structural adjustment program in 1987, the launch of the National Water Saving Program in 1995, the Water Code revision in 2001. But many questions still remain unanswered.

6.1 Securing the Water Supply

6.1.1 Are There New Ways to Improve the National Water Balance?

Despite a number of regional imbalances, partially offset by large transfers, the supply-use blue water budget at national level is relatively well balanced, and should remain so, but for horizons that are getting closer. Otherwise, this balance is very fragile. It is based on a number of dysfunctions: imperfect knowledge of water resource systems, incomplete control of water use, overexploitation of aquifers, outdated and sometimes inadequate institutional system, difficulties to adequately manage the Hydraulic Public Domain, dilapidation of irreplaceable strategic reserves. In the future, due to demographic and economic development, and the increasing living standard of Tunisians, the water balance is expected to experience severe tensions. It is therefore necessary to consider today all available opportunities to improve the blue water budget, both in terms of physical equilibrium and of a better understanding of the various components, or else greater effectiveness of institutions and mechanisms acting in resource conservation.

From now, any new withdrawal is made within the uncertainty margins:

Some water resources representative values are published in the literature, sometimes without question of their origin and reliability. We will discuss in this respect the case of some hydrologic aspects and estimates, about surface and groundwater resources.

(a) **Knowledge and uncertainty on water resources estimates**: Water resources of Tunisia are estimated at 4.85 km^3/year, of which 2.7 km^3 are surface water and 2.15 km^3 groundwater. These amounts include items that are not homogeneous, and particularly the following not additive terms: (i) 0.6 km^3/year of low concentrated surface water, difficult to mobilize by conventional structures; (ii) 0.32 km^3/year from border basins (0.5 incoming and 0.18 outgoing); (iii) 0.6 km^3/year of weakly renewable groundwater; (iv) 0.1 km^3/year of wadis floods recycled in natural groundwater recharge; (v) 0.4 km^3/year of surface base flow originating from groundwater drainage (terms (iv) and (v) are subject to double counting as hydrologists and hydro geologists still work separately); (vi) 0.46 km^3/year of brackish groundwater, whose salinity exceeds 4 g/l, difficult to upgrade for current uses. It is also instructive (Cf.

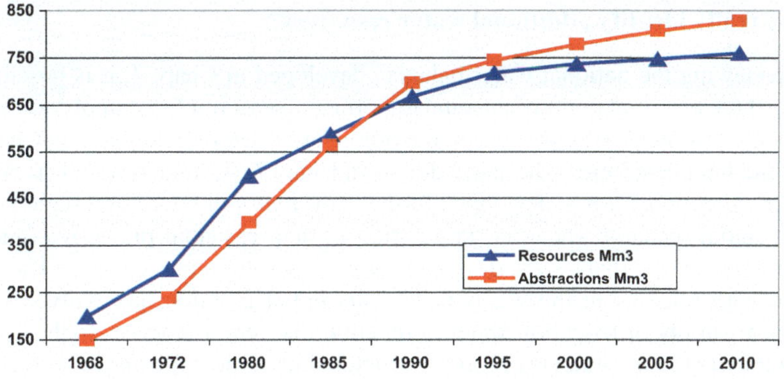

Fig. 6.1 Resources and abstractions of all the shallow aquifers in Tunisia, from 1968 to 2010, million m^3/year (Mm^3/year)

Chap. 4) to observe that the estimation of groundwater resources is constantly evolving, following the level of groundwater withdrawals and the evolution of knowledge, with a recent trend to stabilization.

(b) **On the estimate and mobilization of surface water**: It is currently estimated that 0.6 km^3/year of the surface waters are diffuse runoff, coastal or low concentrated, difficult to mobilize by dams and hill lakes but can be used by water and soil conservation techniques. The value of 2.1 km^3/year setting the limit of easily exploitable surface water was enacted by DGRE in 1990 at the launch of the first 1991–2000 exploitation strategy. Although setting this limit figures as a working hypothesis in the original document, it has been taken as a certainty in all the literature of the past twenty five years. Now, when almost the entire infrastructure of large dams and hill dams has been built, recorded data can be used to clarify all surface water budgets.

(c) **On the estimate of shallow aquifers resources**: For deep aquifers, geological and reservoir knowledge, improved methods for pumping tests and modeling, limited number of boreholes where withdrawals can be counted, permit credible estimates of resources, whereas for shallow aquifers, the great reservoir heterogeneity, the impossibility of counting individual withdrawals from tens of thousands of wells, allow only rough estimates of the resources. This is perfectly illustrated by the graph in Fig. 6.1, where the curve of exploitable resources consistently adheres to that of the inventoried or estimated withdrawals. It is certainly normal for the resource estimate to change and evolve with knowledge progress but the anomaly lies in the fact that groundwater withdrawals are themselves poorly understood. Faced with frequent shallow aquifers overexploitation, this is one of the main weaknesses of the groundwater management which is due to lack of precise studies that the water Authority might invoke to oppose users of the concerned aquifers.

Can we really identify additional water resources?

After reviewing the national water balance, developed in Chap. 4, it is possible to identify three potential pools of unequal importance, susceptible to supply additional exploitable resources: (i) green water, the promising reservoir addressed in Chaps. 1 and 4, and later developed with more details in Chap. 7; (ii) underground storage of northern exceptional flows, developed further; (iii) return of the central region dams to their initial recharge vocation: in addition to their essential protective function against floods, these dams were designed for groundwater recharge of the Kairouan aquifer, with releases in the wadis beds. This initial function has evolved into a permanent supply of irrigation water, a function that requires a very high security, so that the operating strategies of these structures leave no opportunity to recharge by wadis, except in exceptionally wet years. Simulations on long series of dam operations show that a more rigorous water supply to the irrigated area by the dam and priority given to recharge would result in a significant improvement of the water balance of the region, practically reducing to zero evaporation and other losses.

6.1.2 Underground Storage of the Northern Water Surpluses

Artificial recharge, exceptional runoff and storage structures:

Initiated in the early 1970s, the groundwater artificial recharge cumulated volume reaches today, in 2017, almost 1000 million m^3 in Tunisia, which is considerable. However, recharging the single Kairouan aquifer, by artificial floods from Sidi Saad and El Haouareb dams, uses on average two-thirds of these amounts, in order to partly overcome the recharge deficit due to the construction of these dams.

Other artificial recharge projects, where several dozen sites have been developed, did they account for as much successes and substantially increase contributions to groundwater resources? Moreover, what is Tunisia's potential in terms of artificial recharge to aquifers, as a technical alternative to compensate for groundwater over-exploitation? Finally, can we consider storing underground the excess runoff during exceptionally wet years? These are the considerations that deserve to guide a national strategy for artificial groundwater recharge. The third option which, if successful, would offer additional resources, judged to be worthwhile entails a feasibility study of temporary storage on the surface and underground storage of excessive amounts of surface water: Hydrologists estimate that every 10 years, a runoff volume of 2.5 km^3 is discharged into the sea or evaporated on floodplains. Is there a way to recover part of this volume, to transfer it and then inject and store it into suitable underground reservoirs? At what cost and with what environmental impact?

The infrastructure of large dams has been designed to ensure a degree of inter-annual regulation, especially in dry years, and real-time control of large structures allows managing wet years flows, thereby reducing risk and damage associated with major floods. But the capacity of large dams is not able to manage all exceptional

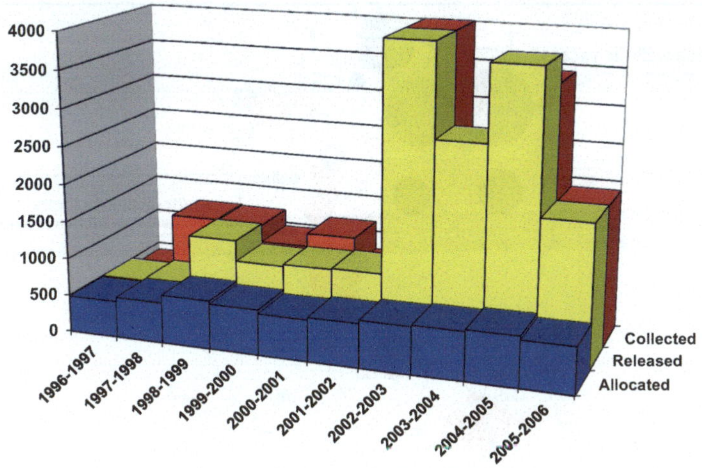

Fig. 6.2 Great dams balance in million m^3, from 1996 to 2006. (The large dams database, DGBGTH)

floods, and most of these exceptional contributions will participate to flood lowlands and join the sea.

The wet decennial provision of Tunisia is estimated (Frigui 2008) at 4.7 km^3/year. The current surface water mobilization capacity equals 2.2 km^3. Uncontrolled inputs therefore represent a volume of 2.5 km^3, available once every ten years, representing a very good chemical water quality. Although these quantities are necessary for the regeneration of coastal ecosystems, sebkhas and other wetlands, it is not absurd to think about transferring part of these quantities to meet human population needs. Figure 6.2 illustrates this situation: volumes discharged by large dams are considerable during wet years compared to those actually allocated. It should be reasonably possible to transfer these significant quantities to deficit areas, through appropriate management based on North Dam interconnection with those of the Centre in particular.

Some preliminary estimates show the technical feasibility of water transfer from Sidi Salem Dam to the Centre (Chekir 2004). Transfers are also possible to the North East (Chekir 2006) through a better management of the Medjerda—Cap Bon canal capacity during the winter season. Such estimates, which deal with a volume of 50 million m^3/year for safe drinking water or to support irrigation, show that such ideas would be considered.

At Cap Bon and Central Tunisia, exist major aquifers whose dry parts (unsaturated zone) would be able to store large amounts of water, which can then be pumped over periods of several successive years. Thus, it was shown (Besbes 1975), that a volume of nearly 200 million m^3, injected upstream of the Kairouan plain in a few months, is "trapped" in the aquifer for decades: 30 years after the injection, losses in the downstream Sebkhas are still negligible. This holding capacity, combined with the importance of unsaturated available volume, make the basement of the Kairouan

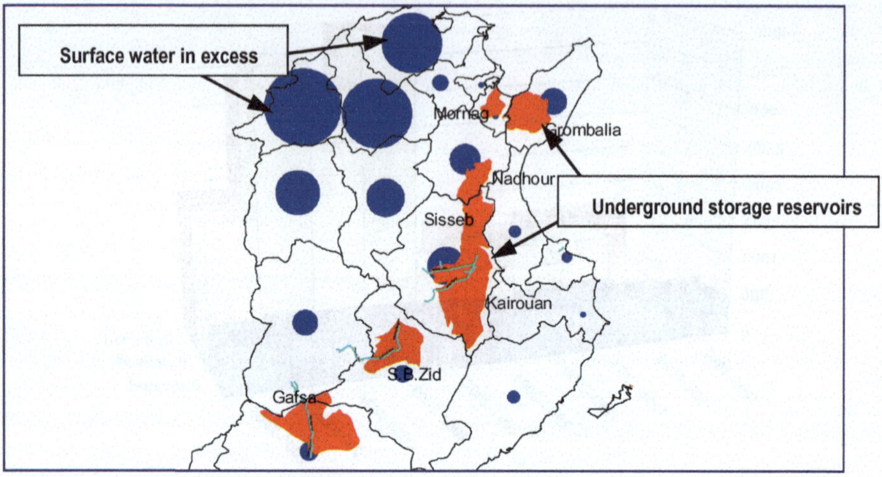

Fig. 6.3 Surface water sources and underground storage reservoirs

plain as a perfect underground storage tank, capable of storing huge amounts of water, a reserve of more than 2 km^3 as a first estimate. Figure 6.3 shows the situation of the greatest potential underground storage reservoirs in the center of Tunisia, with regard to sources of excess surface water in the North.

Can we then think to allocate part of the northern surplus water to route it to underground reservoirs of the Centre; and in this case: (i) are there still some sites available for new dams in the North? (ii) can we precise the regime of flows to such dams and their management? (iii) the North-Centre transfer is it technically efficient? (iv) what are the hydrodynamic and chemical properties of the host reservoirs? (v) what will be the performance of storage, its durability and what would be the effect of climate change? (vi) the North Centre interconnection of existing dams would it be able to fulfill similar functions? (vii) what would ultimately be the economic viability of such a project? (viii) what would be the environmental impact at various levels? All these questions deserve deepening.

But the northern surpluses underground storage is it really beneficial?

There is already a lively debate on this project, and just go to the Kairouan region to achieve the level of expectation vis-à-vis northern waters, a project that farmers consider already acquired and accounted in their medium-term strategies.

On the other hand, opponents of the project consider more useful than those excess volumes be used as environmental water. Indeed, the overall environmental impact for wadis already regulated by existing dams on wetlands and the coastal area of such a locking of the hydrological system could lead to an "ecological catastrophe". Furthermore, the considered exceptional floods occur in the order of days or weeks time scales. To transfer a portion to the Centre, it would be necessary to regulate these floods by new dams and additional storage capacity, that fills and empties once every ten years. This poses a number of questions about profitability and economic efficiency of such a project.

6.1.3 *The Internal Security of Water Resources and Facilities*

Security of the drinking-water supply system:

This section describes how the national water system may be vulnerable and subject to involuntary or deliberate damage. Issues relating to the water safety, especially drinking-water, including securing a major strategic issue, have multiple dimensions:

- Critical situations with lack of resources in the event of drought or circumstantial malfunction for example, or even in case of floods engendering supply disruptions;
- Accidental pollution or poorly controlled contamination;
- Malicious acts of criminal or terrorist nature, or situation of extreme social unrest, or major conflict.

We intend to highlight this last component, and also address the other dimensions mentioned, which are closely intertwined.

Indeed, Tunisia knew more than half a century of hard work and sacrifice to build a modern and efficient national water system, whatever its shortcomings by elsewhere. This system, whose genesis and implementation are described in Chap. 3, has multiple components: mobilization structures, reservoirs, pumping stations and treatment plants, supply and distribution networks…

It is probably the drinking water transport network, the SONEDE network with a total length of more than 50,000 km (20% for water supply, 80% for distribution), which is the most vulnerable. Besides the classic breaks on distribution networks, damages of another kind have been observed after the 2011 Revolution, which took the target of water supply pipes, caused by indigenous groups opposed to inter basin transfers or considering themselves not served by these transfers. Such conflicts have, ultimately, been very limited in number, have not caused significant shortages, and were usually settled amicably, through negotiation and consideration of complaints from people concerned where justified or through the courts in case of undue damage or proven criminal acts.

As for drinking water treatment plants, they are few: twelve throughout Tunisia, including those of Ghdir El Golla and Belli supplying a large share of urban areas, regarded of strategic importance. All these stations are relatively well protected. However, it should not be excluded that there is a risk of malicious acts at this level, which concentrates the vast majority of drinking water in the country. The treatment station is the most sensitive point of the network: after the station, water circulates without further processing to the consumer's tap. At the level of treated water storage tanks, vulnerability is as great if not greater than in the treatment plant. Very small amounts of toxic chemicals introduced at this level, biological agents or toxins, can cause, if not a serious danger to public health, a general panic and a negative psychological impact in important segments of the population. Increased vigilance is required at stations to prevent any intentional act. Several routes should be preferred for this purpose: (i) study the advisability of chemical and biological barriers, capable of causing a rapid reduction in levels of undesirable products; (ii) the development of early detection systems on all sensitive points of the network, and consequently strong monitoring capabilities; (iii) a preventive and effective protection against cyber attacks on SONEDE's computing systems. A similar protocol is applicable in major reserves (large reservoirs and dams), especially in the Medjerda Sidi Salem dam, the keystone of the entire national water system.

Security and the risk culture

Tunisia does not have an exhaustive emergency plan for the management of major risks, such as accidental pollution on collective water supply systems. The 91-39 and 96-29 laws (JORT 1996), and decree 93-942 (JORT 1993), deal with management of natural calamities and disasters (fires, floods, earthquakes, storms) and the national emergency response to marine pollution program, but not with accidental pollution on inland waters and water systems destined to human consumption. To this end and to coordinate all preventive and curative measures in case of major risk or disaster threatening the quality of drinking water, it is necessary to design and implement an ORSEC (Organization of the Help) plan for drinking-water in the existing management procedures for accidental disasters.

Floods or droughts that frequently affect Tunisia bring the notion of risk to the fore in the news. In fact, of all the situations that may occur and involving hydraulic structures, failure of dams, transfer pipes or flood-protecting dikes present the highest risk of human, economic and environmental damage. The requirements for dam safety are extreme. They are omnipresent in all project phases: planning, design, construction and operation. Whether it is flood security, earthquake safety, foundations stability risks: the risk culture is part of dam engineering. International experience shows that large dams are less problematic than one might suppose, while we observe the true risks both on the dikes, whose maintenance protocols are far less precise and rigorous, and on small dams.

Safety of structures and water facilities

Apart from breaking some dikes intended to protect cities against floods and for which the institutional management responsibility is not precisely defined, to date, no particular accidents to dams have been recorded, as they are generally recently built, controlled and managed directly by the Administration or by large public enterprises. In the future, the management situation is set to become more complex, with increasing probabilities of defaults and ruptures, because of the constant increase in the number of medium and small dams, the advancing age of large dams, the latter becoming obsolete and even to certain dangerous despite having scrupulously respected the regulations in force at the construction time.

In this context, it is important that legislative and organizational arrangements need to be put in place, consisting of:

- a permanent technical committee for major hydraulic structures, whose mission is to issue an opinion on any matter regarding safety of the structures, whether in the design, construction, restoration or replacement of large dams and large pipes, flooding protection dikes;
- preparing specific contingency plans for hydraulic structures with great risk, which determine the overall arrangements for the securing of people and goods.

Strategic reserves to secure water supply to metropolises

A limited number of aquifers in the immediate vicinity of the large metropolises should be managed as rescue strategic resources for drinking water to supply populations in case of ultimate crisis, where the majority of water resources and infrastructures have been made unavailable. In the immediate vicinity of each large metropolis, there are one or more aquifers that should be kept in a good quantity and quality, as a “strategic reserve for drinking water.”

It so happens that precisely all such urban peripheral aquifers are depleted, and their water quality continues to deteriorate due to intensive agricultural use, and likely to be useless if ever it should be needed as a source of emergency drinking water. These aquifers include:

- Aquifers of Mornag and Manouba in Greater Tunis, both dramatically overused, especially the Mornag aquifer, where are observed already unacceptable levels of nitrates concentrations;
- The Zeramdine Beni Hassen overexploited aquifer, a freshwater enclave in the heart of the Sahel, the most populated region of Tunisia;
- The Chaffar aquifer near Sfax, also exploited for agricultural purposes and overdraft, whose quality is deteriorating.

All these freshwater supplies must receive special and exceptional status, which allows first to restore their resources in a sustainable quantitative and qualitative status, then save them effectively.

6.1.4 Coping with Groundwater Overexploitation

The need for a new groundwater management mode

Today it is estimated that across Tunisia, groundwater abstractions represent 93% of exploitable aquifers resources, and that these amounts provide nearly 70% of the nation’s water needs. This shows the urgency and importance that must be given to the protection and conservation of groundwater, including the search for more relevant management modes where users are associated to the future of the resource. It must be emphasized, however, that very few cases worldwide can be cited as successful experiences of participatory management of overexploited aquifers.

The Tunisian Water Code had already introduced elements of sound management, but essentially coercive, of groundwater: (i) groundwater belongs to DPH; (ii) obligation to report completed wells; (iii) transformation of water property rights into rights to use water; (iv) fee on withdrawals; (v) establishment of safety perimeters and blackout perimeters. The organizational form involving users in groundwater management follows a number of guiding principles (Ollagnon 1991): (i) the over-exploitation of aquifers highlights the limitations of the concept of common good: each user finds his individual interest to collect ever more on a shared resource; to prevent abuse, we must encourage negotiation process; (ii) without collective choice, the aquifer status is delivered to the conscience of each, which is still insufficient; (iii) in a context of scarcity, the dilemma is to mediate between the aquifer preservation and the economic development; (iv) for groundwater, it is necessary to consider the ignorance, which frees users from any long-term responsibility; (v) the participation of all is possible if the solidarity sense is supported by a strong awareness and a clear objective of maintaining a common good.

There are two ways to seal such awareness: (i) eradication of ignorance effects through education, (ii) introduction of a charge on withdrawals that expresses that water has a value.

Is groundwater participatory management a panacea?

The National Program "Participatory management of groundwater overexploitation", which has major difficulties to be implemented, but is nonetheless part of projects of DGRE, should help to develop the elements for a management strategy which users are the central actors. This strategy is based on the fundamental principles outlined above.

The main objective is to provide farmers, groundwater users, with information, knowledge and skills that enable them to ensure their own sustainable management of groundwater and control their own demand.

The fundamental assumption is that access to data and to scientific knowledge on the aquifer should provide farmers with the ability to make the appropriate choices and take the right decisions for active management of the aquifer, groundwater exploitation over the long term, and sparing use of water resources. This assumption applies on the field by three main actions categories:

(i) Demystifying hydrogeology and water science and teach farmers to take in charge data collection for understanding the aquifer functioning; a precedent exists in the successful implementation of these protocols in the state of Andhra Pradesh in south eastern India, (World Bank 2009);
(ii) Helping to acquire an "aquifer consciousness" to perceive the passage of the freehold status of groundwater to the sense of common good, and to gauge the risk occurred by depleting the water table for the whole community;
(iii) Strengthen local water management institutions and support the emergence of groundwater management institutions at the aquifer level.

Urgency of issues, novelty of solutions, and lack of world examples too, caused a certain excitement among water actors and between experts. Until this year of 2017,

Fig. 6.4 The Bsissi aquifer users' association office (© Besbes, 2008)

there is a strong debate on the order and priorities to be adopted in implementing the actions listed above: should we begin by training users to groundwater mysteries in order to promote the aquifer consciousness development, or should we favor the creation of aquifer management institutions first?

The approach to be adopted is certainly pragmatic: not plotting theoretical schemes on such complex realities, but finding the best solution for each context, physical and social. In practice, two aquifer systems combine the next opportunity and feasibility conditions for participatory groundwater management: (i) a very high level of groundwater threats, combined with a good aquifer knowledge, (ii) awareness, and willingness of local actors to take over the aquifer destiny. The two mentioned cases are those of Bsissi aquifer in the north of Gabes and Mornag aquifer near Tunis.

The Bsissi aquifer, in the north of Gabès: In order to preserve the heavily-exploited coastal area of Bsissi, some 275 boreholes owners, exploiting 1600 hectares of irrigated land, created in 2000 the Bsissi Aquifer Monitoring and Development Group. In agreement with the water authority, the association has obtained to fill illegal boreholes, stop drilling new holes, and benefit of water-saving subsidies. The Group is also responsible for regulating the allocated pumping flows and monitoring the aquifer levels. Considered as a success, this experience is nevertheless isolated in Tunisia and fragile, despite the acuity of the problem throughout the country, and a recent awareness of the phenomenon seriousness, especially by users (Fig. 6.4).

The Mornag aquifer, near Tunis: Groundwater overdraft is acknowledged by all parties, and it is generally agreed that coercive measures advocated by the current regulations remain largely irrelevant. The objective of Participatory management of the aquifer is to develop the elements for a strategy where users are the central actors, to integrate decision-making at the local level, including technical, regulatory, cognitive and institutional aspects. The project is based on the cognitive developments, first and foremost on farmers training, by simplifying hydrogeology and water sciences in groundwater classes. About 1900 farmers are using the aquifer, exploiting 12,000 hectares of irrigated land. Due to the size of the aquifer and its population, the organization of aquifer into sectors is justified during a transition time, where users who have close relationships face the same problems and look for joint solutions. The aquifer consciousness is thus reinforced by common objectives. Over the medium and long term, interactions between sectors become preponderant and require looking for an organization model federating the sectors.

The abstractions fee: an effective tool against overexploitation

The abstractions fee is one of economic tools for water demand management. It already constitutes incitement to water savings prescribed by the Water Code (art. 63 and 53), but the current recovery capabilities of BIRH, the competent institution, does not allow it to properly carry out this mission.

In spite of the reassessment by a joint order of Minister of Finance and Minister of Agriculture (November 3rd, 2014), the rate of the fee remains very low: 5 millimes[1] (mil) per cubic meter of water allowed for agricultural use, and 50 mil/m^3 for uses other than agricultural. This rate does not incite to save groundwater when other water sources are made available. Indeed and for the case of a farmer, the average supply price by surface water irrigation networks (100 mil/m^3) is 20 times upper to the rate of the fee. For industrial and tourism uses, the price of water supplied by SONEDE (1350 mil/m^3 in 2017) is 27 times the fee value. Certainly, the rate of the fee does not represent the real groundwater cost, because we need add investments and functioning costs, but nevertheless and for various reasons, including a total autonomy, as well farmers as industrialists favour autonomous on-site groundwater abstractions.

Elements for a national water use database

Unable to install water meters on the 100,000 producing dug wells and some 10,000 deep wells (much more if unauthorized and illicit drilling wells could be included), it will be necessary to use indirect counting and contradictory procedures to identify hopefully with some reliability the amounts abstracted on groundwater. These processes include irrigated areas analysis by satellite images survey, but there is another method, more systematic, "the water uses census."

Indeed, and like the water balance assessed and published for each water resource unit identified (groundwater, river, dam, hill reservoir), we must identify all the demand of elementary units (district, irrigated public perimeter, private irrigated perimeter, hotel, factory …) aggregated by entity (delegation, governorate) and establish the demand with the same frequency as for the resource assessments. After confrontation, these balances should be regularly published as Water Uses directories.

[1]The millime(mil) is the thousandth of the Tunisian Dinar(DT), 1 DT=US$0.55 on 31 Oct 2014.

Such transparency will introduce greater precision in the resource estimate. This is the process used by DGRE regards the offer: directories for rainfall, runoff, deep and shallow aquifer abstractions, have become irreplaceable as studies and planning tools, and a national references, even if the procedure and methodology are still perfectible. Symmetrically, the Statistical Yearbook of water uses is to design, supply and fit. For example, the United States publish the water used quantities identified by types of use, aggregated by county and state, with a five years interval since already 60 years: the first inventory was published in 1950 (USGS 2013; Kenny et al. 2009).

6.1.5 Securing Water Quality

In the late 60s, Algeria, Tunisia and other countries in search of industrialization, foreign companies investments and jobs creation relocations were probably ready to "buy pollution" according to the quip attributed to President Boumediene (Baudelle 1999) at foreign investors. But beyond the wit somewhat provocative, that famously proved significant of all a state of mind, still alive, on the importance and the priority we should give to preserving the quality of water, either vis-à-vis the drinking water or natural freshwater. In this context, the next two indicators measure the distance that separates us from a peaceful relation, efficient and transparent, with issues related to water quality:

- Thirty five years after its publication by INNORPI, in 1983, the Tunisian standard on the quality of drinking water: NT 09.14, is still not approved, and no drinking water analysis is published in the whole country;
- Twenty six years after the creation of the Ministry of Environment, monitoring and control networks, observation and analysis of water resources quality, are not yet organized in a clear manner, a transparent and effective way between this department and that of Agriculture. In general, natural waters quality does not benefit from investments, human and material, enough to ensure significant knowledge advancement. This remains, in spite of efforts from the National Environment Protection Agency (ANPE and Aquapole 2008), on the initiative of individual researchers without continuity programs, which does not favor the development of a clear vision at the national level.

On the control of drinking water quality

Standard NT 09.14, inspired by the WHO guidelines, had the merit to determine, with some precocity, the elements conditioning water drinkability in Tunisia. Thirty five years after its release, this document is being revised, particularly to take into account the latest international developments in terms of drinkability. To anticipate application difficulties, this standard specified: "… when it is impossible to supply consumers with water that meets the above standards, we should … try to get closer to…during this transitional period this water can be distributed".

Despite this reservation, the health authorities did not wish to make this mandatory officially approving the text: strictly speaking, the consumer therefore has no guarantee that tap water corresponds to the limits set by this standard. In this context, quite informal, the Tunisian consumer relies on its trust in the omnipotence and the competence of both institutions operating in this field, namely the Directorate of Environmental Hygiene and Environment Protection (DHMPE) and National Society for Water Exploitation and Distribution (SONEDE).

(a) ***The Directorate of Environmental Hygiene and Environment Protection (DHMPE)***: Under the authority of the Ministry of Public Health, the Directorate is responsible in particular for monitoring the quality of drinking waters and thermal waters, control of environmental protection and the fight against pollution. To this end, the Regional Hygiene Services levy on the entire country, samples of water in all the transit points where water is intended for drinking: outputs of treatment plants, reservoirs, distribution network and up at tap. If an anomaly is detected, the health authorities immediately warn the party concerned, which shall make appropriate provision and makes corrections.

All communication protocol is confidential, what is legitimate to avoid panic movement or malicious exploitation of a transient and without risk anomaly. What is less so, however, is that the results of analyses are not available to the public. Drinking water is a food per excellence, and it seems to be the right of everyone to have access to regular knowledge on its composition.

Such could calm any irrational feeling of insecurity, illustrated by the surge in bottled water consumption that took over a major part of the population. Consumption of packaged water in Tunisia has risen indeed from 7 L/capita/year in 1990 to nearly 110 L/capita/year in 2013 (ONTH 2013).

(b) ***The National Company for Water Exploitation and Distribution (SONEDE)***: Among the Tunisian water agencies, this institution acquired the reputation of a good communicator, by establishing an efficient statistical service and publishing since the 70s a precise and useful statistical yearbook, where the reader is able to establish the slightest trace of each m^3 of water produced. However, this cult of detail on quantities contrasts with the lack of information on water quality, except to clarify each year in terms of bacteriology, the number of non-compliant analyzes and the percentage of cases unfit for all of Tunisia, generally below the limit required by the Tunisian standard NT 09.14 and threshold tolerated by WHO.

SONEDE appears technically well equipped to master the problems of physico chemical and bacteriological water quality, but has no regulatory obligation to communicate on these issues and does not publish any results on the quality of drinking water. The consumer, however, is entitled to exactly know the tap water quality, especially as the NT 09.14 standards, referred to by SONEDE could be obsolete in certain securities aspects. In terms of all emerging pollutants, it ignores antibiotics, hormones, certain pesticides, which can be found in reservoirs dams downstream of large cities.

Indeed, the real concerns of SONEDE in terms of quality are related to the total salinity. While the NT 09.14 admits salinity up to 2500 mg/L, which may explain the fact that it is not approved, SONEDE still distribute in some areas, albeit limited, higher salinity water. But the National Plan for Improving Water Quality aims to reduce, ultimately (2035) all water distributed by SONEDE at salinities to 1500 mg/L, and using a very ambitious program on Centre West and South regions, with 17 brackish water desalination plants.

Monitoring water resources quality and pollution

DGRE manages a water quality monitoring network comprising 1300 groundwater points spread over the whole country. The measured parameters are salinity and nitrates. ANPE, for its part, operates a water pollution control network, which includes 60 points spread over several regions. The network of ANPE is a new quality monitoring network, which adds to the existing networks of DGRE, DGBGTH (quality of water retained in large dams), DGACTA (hill lakes water, quality of soils), Ministry of Public Health (bacteriological quality of water and sanitary risks), each with its own objectives and specific parameters. All this information should, later, be interconnected into the National Water Information System SINEAU. But interconnecting data is not enough to cover all the issues, the collection of information being still designed and coordinated at a national level to hope overcome any problems. Thus, by way of example, it should be recalled that under the current system, one of the biggest water pollution sites is neither followed nor even listed: this is the Lake El Borma where are rejected for over forty years drilling oil recovery waters of El Borma field (Shel 1988). These amounts, accumulated to nearly 100 million m^3, could constitute a major risk, certainly in the long term for downstream aquifers of Nefzaoua and Jerid.

6.1.6 The Good Status of Water and Human Health

In spite of efforts to establish means to fight against pollution in: (i) water environments and treated wastewater recycling, (ii) drinking water and sanitation systems, problems in relation to human health persist. If in urban areas supported by SONEDE and ONAS, progress is tangible, this is not true into the rural world by universal access to safe drinking water and adequate sanitation for all citizens.

On another level, poor sanitation, improper waste treatment, unsafe disposal methods for industrial chemical products, often abusive use of fertilizers and pesticides in agriculture, sometimes inappropriate water management, all have a negative impact on water quality and indirectly threaten human health. Environmental degradation and the resulting potential health risk represent a trend that could take decades to reverse.

At the water quality level available for human consumption, control and disinfection operations carried out by the regional services of the Ministry of Public Health on collective water points have certainly improved the overall situation. But the water quality remains below the standards required, particularly for wells and water harvesting tanks for family use (SONEDE 2000).

Regarding the population behavior, respect for the protection of water sources areas, transportation conditions and drinking water conservation modes remain problematic in disadvantaged urban areas or rural areas. From this point of view, it is essential to define a protection zone for each public distribution point. At this level, the GDAs responsible of drinking water points and the beneficiary population must acquire the basic rules of hygiene for water. Moreover, in health centers, schools and other public institutions, educational messages with the theme of self-healthy and unsafe water-related disease need to be developed.

These findings lead us to think about building the capacity of public health staff responsible for supervising population and which must have the necessary skills, both technical and educational to convince people of the necessity of compliance hygiene, on one hand, and the importance of the means used, on the other.

The enhancement of water environmental impact in rural areas includes improvements of sanitary technologies. The discharge of domestic and collective wastewater into the natural environment, often in close proximity to housing and the use of cesspools for wastewater, especially when the water table is not deep enough, are a major concern of public health. To overcome this problem, and pending the implementation of a national strategy for rural sanitation, we can consider widespread use of septic tank management actions standardized in rural communities deserved by SONEDE. These actions include awareness raising concern of users, technical support for the implementation, and financial incentives to promote the operation.

Another aspect directly related to human health is the reuse of treated wastewater for agricultural purposes. A regulation exists in Tunisia which defines the quality of treated wastewater to be used, the nature of crops to be irrigated and determines the terms and conditions of use without major risks. Compliance with this regulation is often not very strict, particularly regarding the quality of the produced treated wastewater and direct manipulation of water by irrigators.

Given the fragility of water environment and importance of human health issue, it is essential to promote at all appropriate levels, both locally and in the regional context, the protection of health, both for individuals and collective populations, improving water management, including protection of water resources and aquatic ecosystems, and work to prevent, fight and reduce diseases related to water. The current situation in Tunisia lacks visibility on various stakeholders role particularly in terms of disease control at water resources and uses levels, which defers the whole weight of the health issue to the Ministry of Public Health.

It is therefore important that a clear national vision is established to improve the water health aspects, including in particular the following items: (i) to involve the health protection promotion of environment and human well being by bringing together ecosystems' managers, sanitation and health water professionals in a clear accountability framework; (ii) to develop a holistic framework to address the entire chain of causes and effects, from environmental degradation to the health issues related to water; (iii) to protect against pollution water resources used for drinking or related to ecosystems; (iv) to provide an adequate supply of safe drinking water and compliance with concerted national standards; (v) to ensure for the entire population, within realistic time frames, with a sanitation quality, sufficient to protect human health and the environment; (vi) to mobilize human and financial resources needed to avoid deterioration of the existing drinking water coverage, sanitation and water quality control; (vii) to encourage awareness, participation and effective involvement of the public in decision-making processes related to drinking water and sanitation; (viii) to establish effective systems to ensure monitoring of health status, and facilitating interventions in case of episodes or incidents of water-related diseases.

6.2 How to Achieve the Good Water Governance?

The major water problems in Tunisia are generally known and the solutions identified, although their implementation poses challenges in terms of technical, financial and human capacities. However, it appears that institutional aspects are as important as the purely technical aspects which have often been given priority. Furthermore, the population emancipation has made the social dimension and users' participation as mandatory conditions for sustainable development. Then, new challenges appear for Tunisia: while the major part of available resources has been developed, how to organize resource management to meet growing demand and future needs? In a context where competition between sectors and uses will increase, how moderate water demand, guarantee resources conservation and control risks? Beyond the solutions to improve management methods, how anticipate policies capable to meet the new challenges?

The paradigm shift in water managing is not an option but a necessity. It is important for Tunisia to introduce management strategies that promote environmental and heritage value of the resource. This can be achieved by anticipating and controlling risks, to balance needs for maintaining the social value of water, to ensure the best possible use of the resource and its preservation, to implement practical solutions for the development and enhancement of all available water resources.

These goals will require undertaking in-depth governance and water managing reforms:

(i) Institutional reforms: intersectoral coordination and improving decision making mechanisms, local water management procedures, defining the territorial unity for water management, strengthening and modernization of major public institutions to support water policies;
(ii) Legislative reform with the necessary and urgent revision of the Water Code and the adaptation of the Tunisian water legislation;
(iii) Better optimization of water use by integrating financial, economic and normative tools;
(iv) Cognitive reforms and better use of water knowledge.

6.2.1 *The Institutional Reforms*

Institutional organization and decision making mechanisms

The analysis of the Tunisian water policy components, developed in the previous chapters (particularly in Chap. 3), have clarified the water resources administration, planning, managing, and their characteristics. Tunisia has set the legal framework and set up the administrative bodies and participation procedures at national, regional and local levels. Water resources' planning is organized around "Regional Master Plans for Water Use", but is part of a national vision integrating the whole country. It is a specificity of the Tunisian context, and any important water governance reform cannot escape the State role in financing major infrastructures and in the overall management policy of a limited water resource.

The Tunisian legislator has distinguished between operational decisions and water policy decisions; he also provided advisory institutions which contribute to the development of these decisions. However, the National Water Council hasn't been highly effective, and the Administration has failed in practice to establish a genuine consultation on water policy at a national level. Similarly, the legislator distinguished between the operational decisions from decisions that fall within arbitration and conciliation of water uses when defining Prohibition or Safeguard Perimeters versus Development and Water Use Perimeters (DWUP). While Prohibition or Safeguard Perimeters expresses a technical opinion from the Water Public Domain Commission, reporting DWUP requires water allocation plans, that mean arbitration to reconcile the different uses. In practice no a DWUP has been declared. Here again, we observe the difficulty that actors at different levels meet to develop and implement mechanisms for decision-making which involve some form of consultation and coordination in order to reconcile the uses and appropriately distribute the resource: (i) at the central level where the administration has the upper hand in the development and implementation of water policy and resource planning; (ii) at the local level due to the total lack of consultation mechanisms.

Inter sectoral coordination and water institutions

Any institutional action in water field must, in the future, aim the establishment of effective coordination of the state's action by strengthening consultation and promoting the involvement of public and private stakeholders, in order to improve decision making mechanisms. To enroll in a process that promotes adaptation of the current institutional framework, the implementation of institutional reforms should not undermine the power and authority of ministerial departments and their services. It can, however, attempt to strengthen the role of existing instances by adapting their structures and statutes and reviewing their missions and functions. The actions to be undertaken may target a number of priority functions or institutions that are likely to improve intersectoral coordination in water resources management.

(a) **Establish a permanent high-level coordination, the Higher Water Council**: The coordination role should go beyond mere consultation to encourage the various stakeholders at national and regional levels to work together and harmonize their actions in the framework of a national water policy. These considerations should lead to examine the opportunity to create a high-level body, the "Higher Water Council", assisted by a Permanent Technical Committee, with a possible extension at the regional level.

(b) **Develop an institutional framework for water resources planning**: The water planning purpose is to set basic directions of development and water management at national and regional levels. These can be conceived as part of a "National Water Plan" and "Regional Resources Management Plans" developed by the authorities as part of a recognized national policy that draws its legitimacy from the democratic functioning of advisory bodies and participation. At all levels of planning, development of water resources management schemes can be conceived as part of DWUP under the Water Code. The development and water management at local scales has to comply with indications of plans and schemes on a larger scale.

(c) **Establish perimeters of water development and use at relevant scales**: Resource management within DWUP can be considered at different levels: a river tributary or an aquifer, throughout the watershed, etc. To be effective, the management within DWUP must meet two principles: (i) secure the support of users and involve them in decision-making; (ii) implement appropriate technical and organizational means to drive the Water Resources Allocation Plan and to ensure monitoring and control.

(d) **Promote opening the water sector to private investment and action**: The aim is to relieve the state departments of technical tasks (studies, works, management structures…) so they can devote to a regulation role that ensures the preservation of fundamental balances. This is basically to identify the areas where state intervention is to maintain and even strengthen, and areas where decentralized mechanisms can be more effective.

(e) **Consider risk as an essential component of water resources managing**: The prospects for structural or cyclical deficiencies require strategies development for prevention and intervention measures. They also require arbitration and dispute resolution mechanisms to reconcile the uses and maintain the socio-economic value of the resource. The risk of droughts in particular requires a specific management framework and the establishment of heavy capabilities to mitigate their economic and social impacts. Similarly, the flood risk is an important factor in the spatial planning policy which must consider the risk of "hydraulic disorder" and manage it by taking also account of economic and environmental benefits related to water excess.

The hydrologic basin as a territorial unit for water management

In the coming decades, water problems in Tunisia will experience a multitude of persistent and emerging challenges, and it becomes therefore urgent to think about renovating the present water management system to face the new challenges. The reasons for such renovation are numerous: (i) sectoral management of water has reached its limits and is ineffective to protect resources in terms of quantity and quality: pollution, overexploitation, conflicting uses, etc, are today common issues; (ii) the administrative boundaries of governorates, competence limits of CRDA's action, are not taking into account the whole territory of natural water flows and are not appropriate domains to manage water; (iii) decision-making without proper consultation of stakeholders may be a source of conflict between the different water actors; (iv) actual water issues require a comprehensive vision that, to some extent, is lacking in current management conditions.

Many water problems have local or regional nature and should be resolved at these levels; however, the State will still play a central role in the development and implementation of national water policy. Could hydrologic basin management, which is presented everywhere in the world as the optimal framework for the implementation of integrated water resources management, stand as a favorable option for the renovation of the water management framework in Tunisia? Or else the centralized water management, which formed the winning option for the last 50 years, would present as the Tunisian hydraulic exception, definitively embedded in the new inter regional solidarities sealed for a long time by very large inter-basin transfers?

One of the major methodological problems of water management experience in Tunisia has been the difficulty in designing a relevant territorial unit to analyze and manage the long-term balance questions between resource and demand. The operational management of water problems has been seemingly decentralized and distributed among the main operators such as SONEDE, ONAS, the SECADENORD and CRDA. But in reality, it has been carried out in a fundamentally centralized manner, particularly in planning and financing, and this was considered the best way to optimize the resource valuation at national level: for this purpose the system of large dams and major transfer lines should gradually allow to pool all resources (Northern rivers, Central and Southern aquifers) to interconnect with vast supply networks for drinking water and irrigation. This centralized system was implemented in a progressive way, and paradoxically built in the sixties through an approach where the management units were precisely large watersheds: (i) the use of the three great natural regions, in fact similar to large watersheds, to constitute the relevant territory for hydraulic planning as part of master plans water use, respectively for North, Central and South; (ii) the creation of irrigated areas development agencies, respectively in the Medjerda Valley and Nebhana .

This genesis of the relevant part of regional planning can constitute a model for decentralized management adapted to the territories of the three great natural regions (Cf. Chap. 3, Fig. 3.2): North, Centre and South, combining integrated resource management concepts and water management by basin or sub basins, with the possibility of pooling between the major regions. These larger units are able to offer adequate support to each of the three components of sustainable water resources management: mobilizing the resource, water demand management, water-related ecosystems management. On another level, the proposed management method takes into account all activities that impact on water resources within the territory of the natural flow of water, *the hydrologic watershed*. It also allows considering the basin's capacity to support current and future water demands of different sectors and to obtain a complete vision to preserve uses sustainability.

The basin organism is in this context a space for dialogue, which serves all stakeholders and water users working within the same basin. Its main task is the planning and coordination of water management-related action. To this end, the achievements under the former master plans of water use must be evaluated and updated as a result of a diagnosis that identifies the major challenges of the geographical entity. The new master plans may serve then as planning tools for decades. The three "basin agencies" would probably also be competent to deal with possibly bilateral or trilateral technical issues in trans-boundary sub-basins with corresponding agencies of neighboring countries.

Consolidate large agencies to support and modernize water policies

As we approach the exploitation limits of the resources, the institutional framework is expected to strengthen the sound management of water. As such, major water agencies should improve their services. The reorganization of the institutional framework should allow the implementation of modern management practices and promote coordination between different agencies to optimize their actions in the context of a coherent vision of the water sector.

Beyond the separation ONAS-SONEDE, which is a Tunisian exception and should be reviewed with confidence, major organizations such as SONEDE, ONAS or SECADENORD, must develop their intervention areas to set up as true technological and expertise poles, by investing in Research and Development, technology expertise and innovation. Other agencies involved in competitive sectors, should phase out their commercial activities for the benefit of private stakeholders and evolve their statutes to the promotion, control, coordination, direction, technology development and training.

A Great Scientific Water Agency

Today, Tunisia has an exceptionally rich water data and information potential, and scientific analytical capabilities dedicated. This context allows setting up a center of excellence, an Agency, to support water policies, where develop applied research, expertise, information management systems, transfer of advanced technologies, and on the work of which is based the government policy. This agency could be an offshoot of the current DGRE, strong of a prestigious scientific history and ambitious prerogatives, enriched by a geological pole that is greatly lacking in its current structure. This is to reform and modernize the DGRE by providing it with the necessary autonomy and the critical mass essential to the pursuit of high level scientific and technical activities: make a big national water agency with geological, scientific and technical basement. This institutional model has been chosen by all the great nations that have managed to organize knowledge based rational management of their water resources, like for example USA, France, Great Britain, Australia, Canada and Germany.

An Agency for the Hydraulic Public Domain Protection

Despite great efforts, the Water Authority did not have the capacity to manage and control the development of water resources observed in recent decades and is today in the practical inability to control the many transgressions observed. This crisis is partly due to the unsuitability of an outdated legislative framework, and for another part to the failure of the control structures, due to the inadequacy of the institutional framework to oversee and enforce regulations on the field: the status of dramatic overexploitation in which are a majority of aquifers manifest the great difficulty of the Administration to monitor the validity of licensed or authorized water uses. Such a situation requires building strong and effective institutions, able to quickly counter the threats faced by the water resource sector. Creating a Mastering Agency to protect and manage the Hydraulic Public Domain (HPD) should play a key role to devote the conservation objectives of the resource. This Agency, which could be an offshoot of the current BIRH, will strengthen the institutional framework by undertaking to ensure the Water Police and collection of abstraction charges in the HPD. It is appropriate in this context to organize coordination and cooperation with the ANPE (National Environment Protection Agency), which is intended for its part to ensure the control and monitoring of water pollution.

6.2.2 The Necessary Revision of the Water Code and its Guidance

Despite all the safeguards and monitoring of water resources found in the current Water Code, more modern concepts of water management are not already listed, such as: integrated water management, unity of the water cycle, environmental value of the resource, risk management, role of water information systems, importance of the processes of participation and decision making. The draft revision of the current Water Code, which has been published for comment (BIRH 2013), consolidates the gains, capitalizes the Tunisian experience and breathes new ambitions for water policy. Developed legislative provisions provide the necessary tools for the establishment of an effective coordination and cooperation between stakeholders and the development of mechanisms, arbitration and dispute resolution. The motivations and principles underpinning the reform project are presented in five points in the preamble of the bill whose main ideas are summarized in Box 6.1.

Box 6.1: Extracts from preamble of the Water Code Revision Project

(1) ***A project as a continuation while assimilating new challenges***:

In the new context, Tunisia must both ensure its water security and preserve its resources. Water is a national asset, a public good and an heritage; the heritage character incorporates an element of responsibility towards future generations. Access to safe water is a fundamental right; drinking water supply and sanitation are priority public services. The water code reform project confirms and reinforces more effectively the role of the State; it reaffirms the principle of optimum water efficiency by extending it to all forms of resources, including green water.

(2) ***An integrated and planned approach which enshrines the hydrological unity of the resource***:

Integrated water resources management must conduct a comprehensive approach that respects the unity of natural hydrological cycle, taking into account socio-economic aspects, environmental considerations and risks associated with the operation of the resource, and introduces the watershed management concept. The high degree of artificial water cycle and significant inter-basin transfers lead to strengthen the legislative planning framework: National Plan for Water Resources; Regional Development Master Plans for Integrated Water Resources as part of the Large Hydraulic Regions; Plan for Development and Use of Water in a local context.

(3) ***A scientific risk approach to protect the resource and water quality***:

The project develops legislative tools and institutional mechanisms needed to prevent long-term degradation of the quality and overexploitation leading to reduction in sustainable quantities. It introduces the concept of "good status" of surface water and groundwater. The water saving is considered one of the most important ways to the development, conservation and rational use of water resources. Work on developing savings, improving the quality and protection of national water resources are a public utility.

(4) ***Devote the economic, social and environmental resource***:

Water is both a consumer good which implies a stronger right of access and a production factor that induces economic value. Economic water uses must ultimately bear the real cost of water to encourage sustainable use. The economic valuation of the resource leads to consider water as a fundamental element of sustainable development for regions and a structuring factor for planning.

(5) ***A responsible participation and institutional tools for good water governance***:

The main objectives of the water policy require the establishment of an effective coordination and cooperation between the different actors, the development of mechanisms, arbitration and conflict resolution by creating a higher court, the Superior Water Council, assisted by a permanent technical committee relayed regionally. The participation of all stakeholders in the resource management process strengthens the sense of responsibility and solidarity among citizens. User participation is a given principle, an essential mean for water resources management.

6.2.3 How to Better Use Water

The Standardization and Control of water uses

Tunisian standards are developed by consensus in technical standardization committees of the National Institute of Standardization and Industrial Property (INNORPI). Approved or registered standards in the field of water relate to: (i) qualitative aspects associated with the conditions of use, discharge or water reuse; (ii) sampling and measurement methods; (iii) technical specifications of hydraulic and sanitary equipments (INNORPI 2007). Under these standards, efficiency of use and the fight against wasting water does not appear as priorities.

We should first point out that there is not yet approved standards on the quality of water intended for human consumption even if a standard not approved, is published since 1983. In contrast, mineral waters are considered packaged food and food producers are obliged to comply with specific standards in accordance with World Health Organization guidelines (WHO 2011): the standards NT09-33 for natural mineral waters and NT09-83 standard for packaged water. It would appear that the constraints on the quality of the resource do not allow the supply of the entire territory with water meeting the drinkability criteria consistent with the guidelines of the World Health Organization (WHO 2011). Compliance of Tunisian standards with WHO guidelines requires a strategy for the long term that should lead to improve the quality of drinking water.

Regarding the conditions of discharge or wastewater reuse, the standard for wastewater discharge into the Hydraulic Public Domain, in the Maritime Public Domain and in wastewater pipes (NT106-02) as well as standard dedicated to treated wastewater reuse for agriculture conditions (NT106-03), have constraints or restrictions sometimes considered excessive, resulting in difficulties of application or which undermine the effectiveness of the regulatory mechanism. For example, the standard relating to discharges (NT106-02) advocates extremely low nutrient concentrations in releases to Hydraulic Public Domain (not exceeding 50 mg/L for nitrate, 1 mg/ L for organic nitrogen and ammonia and 0.05 mg/L for phosphorus). These low concentrations can not be achieved using conventional methods of wastewater treatment, including those equipped with tertiary treatment, so that ultimately, in the absence of compliance with the standards, rejecting treated wastewater in the receiving environment continues to require special authorization from the competent authorities which means a failure in implementing these discharge standards. In general, permissions are not required and releases, not necessarily consistent with these very restrictive standards, are discharged into the water environment.

In general, the Tunisian standards system in the field of water is still insufficient vis-à-vis the desired objectives for a rational water resources management. Building a regulatory and normative system should aim first to support strategies to improve delivery systems and water management. In particular, the standardization of hydraulic equipments and water leakage monitoring procedures (audit, intelligent valves and flow measurements, pressure controllers, instrumentation and software in specialized leak detection …) will support the water efficiency effort and the fight against waste, and make available to various actors the technical and economic information necessary to establish technical specifications that improve the water systems efficiency.

The economic and financial instruments: equity versus efficiency

During the second half of the twentieth century, the essential objectives of the water policy in Tunisia were to equip the country, mainly urban agglomerations and in a lesser extent rural areas, by an operational service of water mobilization, distribution and sanitation. Achieving these goals has achieved an unprecedented level of equipment and services on a significant portion of the territory. Beyond the direct satisfaction of water needs is agriculture which has also received massive hydraulic investments supported in the long run by the State. However, because of the "water stress" situation, the explosion of demand and abstractions on the resource, as well as more or less latent conflicts between sectors of uses, the water was never "easy" in Tunisia and economic aspects of water management were always present in the water policy.

Although it is generally accepted as a public good, essential for everyone's life, regardless of where he lives and the financial means at its disposal, water has been considered, in a more or less explicit manner as a commodity that has a value and a price. So, the question of the price of water is perceived as central, and often causes pressure from users and crystallizes many debates particularly on social and equity aspects, and on the sectoral use and the water allocation efficiency.

Regarding equity issue, the drinking water pricing system adopted by SONEDE is of a progressive type, which aims to encourage water savings with a minimum share price for consumption below 20 m^3 per quarter or a moderate price, in rural regions, for GDA connected to the SONEDE network. In dispersed rural areas, drinking water charges depend on the production cost, very different depending on the situation and often reach much higher values than those made in the urban world. Survival of GDA currently depends on grants by the CRDA in the form of free maintenance interventions or equipments replacement. The gradual agglomeration of populations in the medium and long term and the passage of small villages to the municipal regime is the only way to enjoy a national equalization system for drinking water rates which paradoxically favors currently the urban world.

To enhance equity between rural and urban, and between different GDA, it is important that a rational pricing system or water subsidy can take into account the payment capacity of drinking water by various population categories. The water price question deserves to be opened. It cannot be closed without the water sector national operators (SONEDE, ONAS, SECADENORD, CRDA, GDA) revisit, in conjunction with various partners, their "commercial" services. They must offer to users and to the public a new pricing policy that takes into account the various interests: the interest of operators and management agencies to balance their budget, the interest of water users, the interest of public finances and finally the interest of the national economy which has invested heavily in water projects. This change would result in strong price signals and requires to fully anticipate the impacts in terms of social, economic, environmental and political consequences.

One of the pitfalls in this policy would be to establish a single price or a uniform solution throughout the national territory, and not take into account local realities. Indeed, the configuration of the resource differs from one place to another and from a given usage sector to another one, so that the cost and value of water are not the same in all regions of the country. It is from the achievements of recent decades and local experiences that we can verify the validity of a particular solution. In any case, this historic element should always be in mind: water is a vehicle for social cohesion, sharing and collective responsibility. Furthermore, inter basins waters transfers have created inter regional solidarities that should be preserved.

On another level, the debate has always existed in Tunisia about adoption in water management of certain principles of sectoral efficiency, the level of the allowance that leads to affect water priority to sectors offering the best economic profitability, including drinking water for urban services, tourism and industry rather than agriculture. Furthermore, and in order to improve the intrasectoral efficiency, the water used must be allocated in a special way to productive activities that generate the best economic benefits, such as irrigated production of market gardening and tree crops obtaining a higher price on the domestic and global markets at the expense in some cases of cereal and forage crops at the base of the food and feed.

Progress achieved in improving the efficiency of water are real but still very insufficient, and reforms are needed to improve performance: water conservation programs undertaken, designed to get water existing systems to supply additional users to other activities where economic and social impacts are higher, constitute the beginnings.

It is therefore necessary to develop and strengthen economic analyzes highlighting the financial savings that can be achieved by the implementation of policies aimed at mastering water demand. To this end, the economic information needed to evaluate the cost and effectiveness of water services should be systematically collected. These analyzes should also provide a basis for improving intersectoral water efficiency. The possible gains by a more efficient resources allocation among different uses and sectors (agriculture, industry, tourism ...)

can in effect be evaluated locally—depending on the context and the hydrological value of goods—by "cost-benefit" studies of different options, including the costs and benefits of environmental and social externalities. If some Mediterranean countries begin to determine their allocation trade-offs based on an optimization criterion of type "more value per drop," the optimization studies and allocation for waters with different qualities-salinities are scarce or non-existent (Blinda 2012).

In any case, the solutions trying to preserve the balance efficiency-equity can not be defined on a purely objective basis, but also by a real user participation in water management and negotiation of interests involved in the parties concerned, considering that water is a public good.

6.2.4 Better Valuing the Knowledge on Water

Information systems, modeling and the role of research

The need has never been greater to mobilize science and scientific research to improve our knowledge and how to act on the water, especially to anticipate problems in the future that may be detrimental to the country development and to our daily lives. Our ability to anticipate is based on the development of three key sectors, three issues, whose mastery is as many challenges to better prepare the water sector future in Tunisia: (i) Information Systems, (ii) modeling, and (iii) efficiency of scientific research.

(a) **The information systems and decision processes on water**

The preamble to the XIth Plan for economic and social development of Tunisia 2007–2011 invoked, among others, the importance of foresight, the increase of knowledge, the mastery of technology and changes in immaterial economy. The water sector is fully in this direction: it is important that all players fully assimilate the new water challenges, assume the objectives and mastered the technical means to achieve them. The information and communication, which are the two pillars of cognitive nations then become essential components of strategies for the rationalization of water use.

Application of these new guidelines can be achieved only through the establishment of the National Water Information System (NWIS, or SINEAU). The Ministry of Agriculture, as a department in charge of water, plays a central role in the formulation process of the national water policy. However, the tools which are still used by the Ministry to analyze and transform its information in order to orient this policy are no longer at the level of the exceptional volume of existing data and data acquired every day, nor of the skills and expertise existing in the country.

The water policies processing are classically based on the development of: (i) structured databases; (ii) dynamic and interactive geographical information systems; (iii) powerful modeling tools. The keystone of this device is the establishment of the

National Water Information System that should aim at ensuring the interconnection of databases, geographical information systems and water modeling tools nationally to strengthen the strategic importance of hydrological environments knowledge for better water management. It is in this context that the modernization of the water information system stands today as a must. The Tunisian information system must be aligned with the new strategic challenges of water policy, where information plays a major role.

With the implementation of NWIS, Tunisia will be part of a historical perspective, with an evolution towards establishment of structured information systems, but to succeed, the project must be based on substantial regulations: data acquisition on water resources and water uses, and the implementation and maintenance of information systems on water, both on the technical and institutional plans must occupy the place they deserve in the new Water Code. This modernization, moreover, can not succeed without genuine political will: it is clear that the NWIS project is now in its fifteenth year of studies, design and gestation. The outcome and the true success of such a project is for all of Tunisia a double challenge, a double litmus test:

(i) Sharing public information, and water is a public domain par excellence, is one of the main appearances of modern democracies. With Decree Law 2011-41 (JORT 2011), Tunisia is officially committed to this perspective, but the Tunisian Administration is struggling to adapt to this new text, whose practical application is very difficult to be felt on the field. The actual outcome of NWIS would be a concrete proof of this new direction.
(ii) To communicate knowingly its statistical data to the Public, Water Administration must first rationalize its own systems and collection networks, critical data analysis, and organize consultation practices among its own services and departments, central and regional, to improve the quality and reliability of the information produced. But outside of a few pioneering organizations in this field (INM, DGRE, SONEDE) these practices are still uncommon and very weakly anchored in water services and water agencies.

(b) **The modeling challenge**

The second issue is that of "Modeling": the water sector is a highly complex and multidisciplinary system. Collect data and information from various sources and of various natures, validate, verify their reliability and consistency certainly is a necessary step. Valuing them by developing our capacity to deal with and improve our representation of phenomena makes the implementation of modeling tools able to reproduce past trends in order to explain and predict current status, so anticipate future developments. The model is probably the most powerful mediator, able to develop expertise and inform decision making. Highly developed in universities but in a somewhat disparate manner, unorganized and often disconnected from application sectors, modeling is almost totally absent from decision-making circles in the water sector. Again, political commitment will be a key element to change mentalities: closer, sometimes in spite of themselves, actors from universities and application areas will certainly require much more than a good dose of imagination.

(c) **The role of science and research**

The third challenge is our ability to make science and scientific research instruments that can help us better prepare for the water future in Tunisia. Now what we see is that most research centers and laboratories working in the water sector develop high quality scientific activity. This activity maintains a basic scientific culture and reproduction of academic elites, but confined to this mission, research in the field of water failed to evolve towards new horizons: (i) while drawing real problems, it does not fit in the perspective of major national development projects and does not constitute a force of proposals for decision makers; (ii) it falls, moreover, not always in great federating scientific programs (basic or applied, national or international) liable to induce real benefits for the country. This is a somber observation; but we must recognize that our research results have not yet succeeded to impact the major water issues in Tunisia. The challenge is immense, a challenge that the Tunisian scientific community is able to meet, but at the price of courageous reforms.

The future of water is in the construction of a learning society

Tunisia is characterized by scarce water resources already intensely mobilized. Over the coming decades, a significant water needs increase is predicted, including an increasing demand for irrigated agriculture with worsening groundwater overexploitation. With changing demography, increasing urbanization and the general rise in living standards, the comprehensive water balance indicates a growing dependence vis-à-vis virtual water imports to compensate for food deficits. At the 2050 deadline, and if the water governance modes were not evolving drastically, the water dependency ratio could reach 50%, against 39% in 2010.

However, and despite this disturbing finding, Tunisia remains among the countries in the world where water is not used in the most efficient manner:

(i) The overall irrigation efficiency (performance networks and efficiency to the plot) has stagnated at 60% (while it reached 80% in similar climatic zones in Australia and in the southwest of the United States);
(ii) The overall efficiency of drinking water use (comparing consumed to collected volumes) remains at 72%, despite efforts of SONEDE in terms of preventive maintenance and networks improvement;
(iii) Some pricing systems and subsidies do not always encourage users to save water and give the wrong messages about the real situation of water resources in the country;
(iv) Overexploitation of groundwater does not appear to promise any prospect of short-term solution,
(v) The reuse of treated wastewater remains, for various reasons, very inadequate.

In reality, achieving a higher level of overall efficiency of water systems requires empowerment of the whole society, which implies a massive appropriation of knowledge. It implies first an expertise with user education, and learning water management by children through school education. But there is also access to the whole of society to general knowledge. The "Knowledge Index" developed by the World Bank (2013)

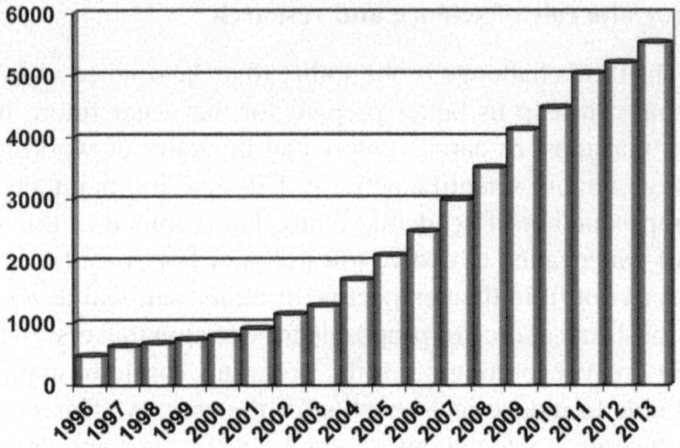

Fig. 6.5 Tunisian publications indexed in the web of Science (1996–2013); (data from MESRS-PASRI 2015)

integrates the country level basic education, business and innovation capacity of the country and access to information and communications technology. At this level of appropriation of knowledge by society, Tunisia is certainly located in the small leading group of African countries, but it is also very badly placed when compared to northern Mediterranean countries.

Another indicator of the society knowledge level is scientific output, where Tunisia, which collects dividends of the 1996 Research Law, is currently in a very good gradient of creativity (Fig. 6.5), which should be able to wear fruit in a few years. Yet this new scientific momentum has not yet given a significant impact on society:

(i) not in weight, visibility and attractiveness of universities, whose world ranking is still largely unacceptable;
(ii) nor that, particularly in the water sector, of involvement in large projects and in guiding public policies, areas, discussions and decisions where researchers and scientists are not guests welcome, nor yet quite comfortable and heavily invested.

The future of water in the country will rest largely on the modernization of the management of all water systems, which implies a massive cognitive evolution. This condition requires a strong emancipation of the whole society, to achieve the objectives of efficiency. It requires that all stakeholders understand water issues, and that the public opinion appropriates the water issue as a project for the future. Information, participation, users' education, schools, scientific research, form the essential components for mastery of knowledge on water and rationalizing its use.

the reforms to be introduced for modernization of data management and information, on local and democratic water management modes, on training users and farmers to water savings and efficient management of resources, on the approach of the water in school education, the development of innovation and research applications, are expected to help stabilize and sustainably improve water security of Tunisia.

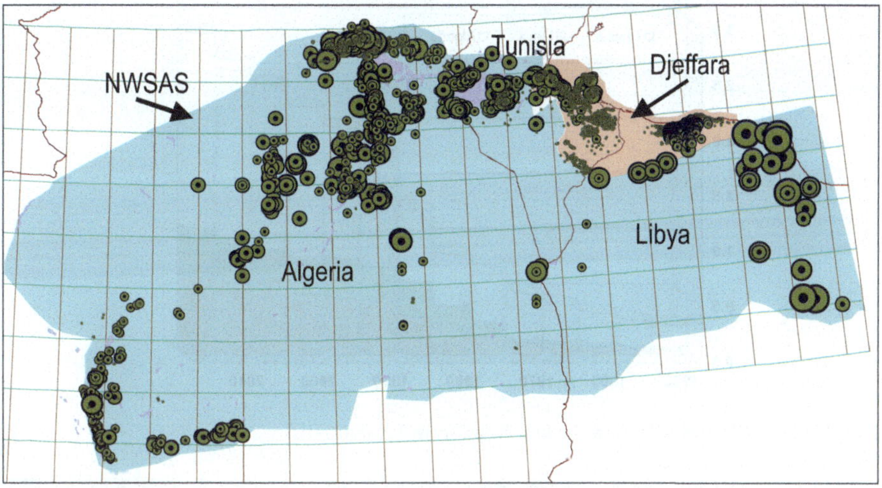

Fig. 6.6 Delimitation of NWSAS, Djeffara, and exploitation boreholes

6.2.5 Sharing Knowledge to Sustainably Manage Transboundary Basins

Tunisia shares two large transboundary aquifers with its neighbors: the North Western Sahara Aquifer System (NWSAS) and the Djeffara Tunisian-Libyan coastal aquifer. The NWSAS covers more than one million km^2 in Algeria, Tunisia and Libya (Fig. 6.6). The layers extension and thickness have enabled the accumulation of considerable reserves, which are weakly renewable and only partially exploitable. Over the past 60 years, NWSAS withdrawals have increased tenfold (Fig. 6.7), reaching three times the system average recharge, and the aquifer now faces several major risks: high transboundary interference, water salinization, artesianism disappearance, outlets depletion, excessive pumping heights. The three countries concerned with the NWSAS future are therefore condemned to seek together a form of responsible management. This was the birth of the "NWSAS Consultation Mechanism", a formal institutional framework for joint management of this shared resource.

The technical problems encountered by the NWSAS countries naturally led them to organize together to conduct a major survey: the partnership practice during the project gradually forged mutual trust between the technical teams, the consciousness that problems encountered by some depend in part on the actions of others, the conviction that joint action increases the effectiveness of solutions, and the certainty that information and data exchange, which forms the basis of all solidarity, have become, with the end of the project, an activity not only possible but necessary and beneficial: the developed database containing all current and past information was operational and made accessible to the three countries.

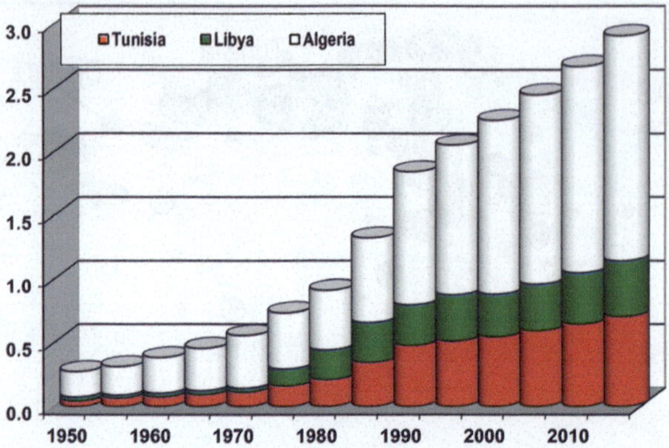

Fig. 6.7 The NWSAS withdrawals, km^3/year, from 1950 to 2015

Fig. 6.8 Official approval of the first consultative structure, at FAO headquarters, Rome, 20 December 2002 (© M. Besbes, 20/12/02)

In this respect, the goodwill of the three water authorities for the communication of information has been exemplary. In addition, the NWSAS Simulation Model was already available and operational in each of the countries. An effective form of consultation consisted first of all in maintaining, developing and constantly updating these tools: Database and Simulation Model. This mission has been entrusted to a permanent body. First approved under the aegis of SSO (2017) and FAO (Fig. 6.8), the Consultation Mechanism was created in 2006 by the Joint Declaration of the three Ministers in charge of Water.

For Tunisia, maintaining the good status of the NWSAS is a major challenge for its water security, in two ways: (i) the NWSAS abstractions provide 30% of water withdrawals required for irrigation in the country, (ii) a sharp water levels decline would lead to a contamination of Chott Djerid over the medium term, which means an irreversible and permanent loss of most of the NWSAS resource. Moreover, each of the three countries would be subjected to a lot of nuisances no less catastrophic and irreparable in the event of massive deterioration of the aquifer status. The well-understood interest of all therefore consists in developing a strong, autonomous and intellectually independent common institution, having a high scientific capacity, able to draw up opinions and recommendations which Member States take into consideration.

Since 2006, the Consultation Mechanism (CM), hosted by the SSO and financed by the three concerned countries, is functioning on a permanent basis. Nevertheless and after ten years of operation, the CM remains an informal and precarious institution. The difficulty of reaching an intergovernmental agreement officially organizing this new institution, the human resources certainly of a very high level but in insufficient number, the irregularity of material means, the absence of texts governing relations with SSO, all these shortcomings are beginning to affect the entrepreneurial capacities of this fragile institution, whose results to date, though considerable, do not seem to be commensurate with its initiators ambition. Despite these weaknesses, the CM must still be considered as a great success, as it is a testimony that a transboundary basin responsible joint management cannot be conceived without a substantial intellectual and material common investment in terms of knowledge: the quantity and quality of data exchanged between countries, the high quality level of the database and of the simulation models put in common, are indeed considered by all observers as a "common pool transboundary water knowledge basin" unique to the World.

References

ANPE and Aquapole (2008) COPEAU; "Réseau de Contrôle de la Pollution des Eaux en Tunisie" LIFE Pays Tiers. http://loire.aquapole.ulg.ac.be:8081/

Baudelle G (1999) Industrie et environnement, entre contrainte et gouvernance. *Revue de géographie de Lyon* 74(3)

Besbes M (1975) Etude hydrogéologique de la Plaine de Kairouan sur modèles mathématiques; Rapp.int.DGRE/CIG Ecole des Mines de Paris; Tunis-Fontainebleau; 1975

BIRH (2013) http://www.semide.tn/CDE_2.pdf. Consulted on 8 Aug 2013

Blinda M (2012) Vers une meilleure efficience de l'utilisation de l'eau en Méditerranée. *Plan Bleu, Cahier* N° 14, Novembre 2012

Chekir H (2004) Possibilité de valorisation des eaux excédentaires de la Medjerda; Transfert Sidi Salem Nebhana; MARH, DGBGTH; 2004

Chekir H (2006) Augmentation de la capacité de transfert du canal Medjerda Cap Bon; MARH, DGBGTH; 2006

Frigui HL (2008) Les évènements extrêmes en Tunisie, sècheresses et inondations; Colloque Gestion des Extrêmes; IME, Tunis 24 Apr 2008

INNORPI (2007) Catalogue 2007 des normes tunisiennes

JORT (1993) Decree No. 93-942 of 26 April 1993, laying down the procedures for drawing up and applying the national plan and the regional plans relating to the fight against calamities, their prevention and the organization of relief. Journal Officiel de la République Tunisienne

JORT (1996) Law n° 96-29 of April 3, 1996, instituting a national plan of urgent intervention to fight against the events of marine pollution. Journal Officiel de la République Tunisienne

JORT (2011) Decree-Law No. 2011-41 of May 26, 2011, on access to administrative documents of public bodies. Journal Officiel de la République Tunisienne

Kenny JF, Barber NL, Hutson SS, Linsey KS, Lovelace JK, Maupin MA (2009) Estimated use of water in the United States in 2005: U.S. Geological Survey Circular 1344, 52 p; http://pubs.usgs.gov/circ/1344/

MESRS-PASRI (2015) Étude bibliométrique sur la recherche tunisienne, 072-GOV-1-3-DPC2/3-DPC2

Ollagnon H (1991) La gestion patrimoniale des eaux souterraines à travers l'exemple de la nappe phréatique d'Alsace - Comité Pollution des nappes phréatiques de l'Académie des Sciences, Paris, janvier 1991

ONTH (2013) Office National du Thermalisme et de l'Hydrothérapie. http://www.thermalisme.nat.tn/site/publish/content/default.asp?lang=fr, consulted on 17 Nov 2013

Shel F (1988) Devenir des eaux de production du gisement pétrolifère d'El Borma; DEA, Fac.Sc. Tunis

SONEDE (2000) Etude du secteur de l'eau potable en milieu rural- Perspectives, Mission 7 : Environnement, santé et aspect social, IGIP/SCET-TUNISIE

SSO (2017) The NWSAS Consultation Mechanism: an instrument for technical and political cooperation on the NWSAS transboundary basin. http://www.oss-online.org/fr/m%C3%A9canisme-de-concertation-du-sass. *The Sahara and Sahel Observatory*

USGS (2013) http://water.usgs.gov/watuse/, consulted on 23 Aug 2013

WHO (2011) Guidelines for drinking-water quality, fourth edition. World Health Organization. http://www.who.int

World Bank (2009) République Tunisienne, Réflexion stratégique sur l'eau potable et l'assainissement en Tunisie - BIRD/Rapport 44744-TN

World Bank (2013) Knowledge index. http://info.worldbank.org/etools/kam2/KAM_page5.asp

Chapter 7
The Holistic Water Balance: Blue Water, Green Water and Virtual Water

Contents

This chapter is devoted to the analysis of the water balance related to food needs of Tunisia. Quantifying in terms of water-equivalent the different contributions to the food demand clarifies the relationship between agricultural policies and water policies, providing valuable elements for decision-making. This chapter comprises five parts:

(i) The first part presents the methodological basis of holistic water balance analysis. This one is based on a conceptualization of the hydrological cycle that identifies two components of the resource: blue water and green water. The

M. Besbes et al., *National Water Security*,
https://doi.org/10.1007/978-3-319-75499-4_7

water footprint concept refers to consumptive use of each of these two kinds of water. The water use of human activities is also composed by indirect use of virtual water related to external trade.

(ii) The second part is devoted to the assessment of green water footprint of the rainfed agricultural production; wherein an overall linear regression model between rainfall and crop yields is developed. Validated by a classical water balance model of the soil-plant system, the model is used to determine the water footprint of cereals and olive productions and then, the green water potential of rainfed agriculture in Tunisia.

(iii) The third, fourth and fifth parts deal respectively with the assessment of the water footprint related to Tunisian agricultural production as a whole: irrigated crops, crop production, animal production and establishes water requirements of food demand.

7.1 Water Footprint, the Keystone of Holistic Water Balance Analysis

7.1.1 Blue and Green Water Management

The inland water resources system is fed by precipitation and, depending on its course, water is referred to as Blue Water (BW) and Green Water (GW). BW forms reserves and flows in continental hydrosystems: surface and underground flows that are directly available (e.g. rivers, lakes, aquifers and groundwater flows). The GW includes soil water, used by forests, grasslands and rainfed crops.

Current water policies are concerned only with exploitable BW resources and much progress has been made in this field. In this regard, Integrated Water Resources Management (IWRM) has become a standard paradigm for freshwater policy development, allowing for taking into account the different components of water resources and their interactions. But IWRM still concerns only BW, and the significant contributions of GW are not taken into account in the analysis of water balances and not considered in the characterization of water stress situation.

The term GW, introduced by Falkenmark (1995), was defined as the fraction of precipitation infiltrated in the root zone and used for the development of biomass, whether of natural or agricultural origin, the BW representing the rest of the resource. The amount of GW is not directly usable in the form of classical water resources, but actually used partly through agriculture for food production.

The distinction between BW and GW proposed by Falkenmark had the merit of making visible the potential of GW. It implies that the management of water resources must not be limited to mobilized water resources allocated to human activities. It must also take into account the water needed to develop biomass. One of the first consequences of this wider perception of water resources has been to rethink water balances and to put into perspective or revisit the criteria that characterize situations of water stress, (Chahed et al. 2008).

This vision of the hydrological cycle and its application to agriculture distinguishes: (i) the productive GW which represents the GW used for terrestrial biomass production systems (crops, forests, grassland), and (ii) the non-productive GW which represents essentially direct evaporation from the soil.

The fundamental implication of expanding the concept of water resources to BW and GW is to consider precipitation as the primary water source instead of the conventional approach that considers continental flow as the primary water source (Besbes et al. 2002; Falkenmark and Rockström 2006). In a more general way, integrated approach constitutes a shift in thinking water resources governance and management, leading to an abundance of new concepts in the fields of water security and food safety. Moreover, this approach is particularly relevant for the arid zone context where both types of water participate effectively in the agro-food balance. A decisive advancement could come from systematic implementation of widened GW-BW perspective within the water management and governance modes, in particular in water scarce regions.

7.1.2 Water Footprint and Virtual Water Trade

Introduced by Hoekstra and Hung (2002) from the University of Delft, the concept of Water Footprint (WF) is developed from the perspective of water consumption by analogy with the concept of ecological footprint. The WF represents the consumptive use of water to produce the goods or services consumed by an individual or a community. The Delft University team contributed greatly to the evolution of the WF concept and to the enrichment of its content, so as it has become widespread and been applied to different products at different regional, national and international scales. Because of its environmental significance, the assessment of the WF of products, processes and organizations has become such an important and precise issue that the International Standard Organization (ISO) has addressed this issue and developed in 2013 a new standard (ISO 14046) to set a framework for standardizing its evaluation.

The WF concept has evolved considerably in line with evolving concepts in water resources. The first assessments of WF were limited to the determination of general WF. Next contributions distinguished between BW and GW, and the most recent introduced grey water (Hoekstra et al. 2011). The BW Footprint (BWF) represents consumptive use from BW sources, and the GW Footprint (GWF) represents the fraction of GW that is consumed by agricultural lands. This corresponds to what Falkenmark and Rockstrom (2006) call productive GW covering crop and permanent pasture areas. The gray water footprint, presented as an indicator of freshwater pollution, measures the degradation of the resource. It represents the amount of fresh water required to assimilate pollutants to meet specific water quality standards. This is a new concept that is very relevant but has not yet been studied as widely as those of BW and GW; its assessment remains somewhat ambiguous and may be incomplete:

(i) on one hand, pollution standards are not the same everywhere and in some cases they do not exist, (ii) on the other hand, this category should include overdrafted groundwater which contribute greatly to water resources degradation.

The total WF of a geographic area (country or region) includes two components: the Internal Water Footprint (IWF) and the External Water Footprint (EWF). The IWF is the part of the water used to produce goods and services produced inside a given area. The EWF represents water used to produce goods and services imported and consumed within the area considered. The EWF corresponds to the net Virtual Water (VW) flux, an innovative concept introduced earlier by Allan (1997, 1998). The VW is defined as the volume of water required to produce a commodity or service, transferred from one place to another, or from a country to another one, as a result of product trade.

The virtual water concept provided the basis for forward interdisciplinary thinking of water that understands the interactions between water resources, water uses, agricultural production as well as the interactions between the economic, political, social and institutional issues. The VW associated with international trade has often been interpreted as a complementary contribution that allows water scarce countries to mitigate the physical lack of water resources for domestic food production. Furthermore imported VW is supposed to reduce the domestic demand for water and this reduction has been viewed as water "saving". VW is thus presented as realistic means that can be used to achieve water security objectives. Recent findings seem to put into question, or at least relativize, these ideas, which have become dominant in connection with VW issue. In fact, a country's VW trade is not in all cases determined by water resources availability or water scarcity: for economic efficiency reasons, some water-rich countries have resort to food import, while other water scarce countries achieve favorable VW trade balances.

7.1.3 Towards a Holistic Approach to Agriculture and Water Management in Tunisia

The programs of mobilization, transfer and water management implemented during the last five decades have led to an almost total development of water resources and a tendency to stabilize water supply. Coping with the expected demand increase, raises the question of the conditions under which water resources are able to ensure national water security in the long term? This question is in relation with the objectives that should be assigned to the agricultural policy, to the development of water resources, to the role of the conservation measures and to the importance of non-conventional water resources production.

Taking advantage of the water program developed in Tunisia, the irrigation sector has marked substantial progress: the total area equipped for irrigation was multiplied by a factor 7 over the past four decades and is developing over 410,000 ha in 2010, about 8% of the total utilized agricultural area. Irrigation is largely practiced for vegetable crops in various parts of the country; and for fruit trees, in particular citrus in the region of Cap Bon, date palm in the oasis of the South and secondarily for cereal crops and fodder in the north and the center of the country. The irrigation sector consumes 75–80% of the total blue water withdrawals (BW) and contributes to one third of the agricultural production value. This means that two-thirds of the agricultural productions come from rainfed agriculture. Nevertheless, the Tunisian strategy of agricultural development gave priority to the irrigation sector, and rainfed agriculture has not received the same attention in terms of investment and has not known comparable progress.

The area cultivated under rainfed conditions (about 4 million hectares) is variable because of random rainfall conditions (Besbes et al. 2014). The two major farming systems, rainfed cereal crops and orchards (mainly olive trees), occupy more than two thirds of arable land in different soil environments, more or less favorable for agricultural intensification. The grain agriculture lies in the north, while olive trees are grown in the central and southern regions. Whilst rainfed agriculture has not registered significant expansion in recent decades, the sector tried to improve its performances: the cereals average yield has more than doubled over the past forty years, and there are still significant margins for further improvements (Rezgui et al. 2000, 2005).

The concept of WF was applied to the analysis of water resources involved in agricultural production in Tunisia. The WF of agrifood products is determined by reporting the volumes of water resources, regardless of their origin (rainwater, irrigation water), to agricultural production. The same approach is applied for the determination of VW flows related to foreign trade of agrifood products. The results obtained were the basis of a series of works which made it possible to identify the different forms of water resources, their nature and origins, and to clarify certain aspects of food security by confronting all water requirements to the total water potential of the country. The analysis of the overall national water balance, initiated and continued in Tunisia over the last fifteen years (Besbes et al. 2002, 2010; Chahed et al. 2008; Chouchane et al. 2013) is part of an approach that is spreading around the world. Several recent studies have indeed been carried out in order to establish precise integral assessments of water budgets in many countries: Morocco (Schyns and Hoekstra 2014), China (Liu and Savenije 2008), India (Kampman et al. 2008), Netherlands (Van Oel et al. 2009), Spain (Aldaya and Llamas 2009), Belgium (Vincent et al. 2011), France (Ercin et al. 2012).

The important contributions of GW in food production are not directly reflected in water resources planning. This is based on a hydraulic approach that considers the potential of BW resources, so that the potential of GW resources associated with rainfed agriculture and rangelands is often not well known. Given the importance of these resources in the food production system, the development of accurate methods for the identification and estimation of their potential, their spatial distribution and variability cover major issues vis-à-vis of food security objectives.

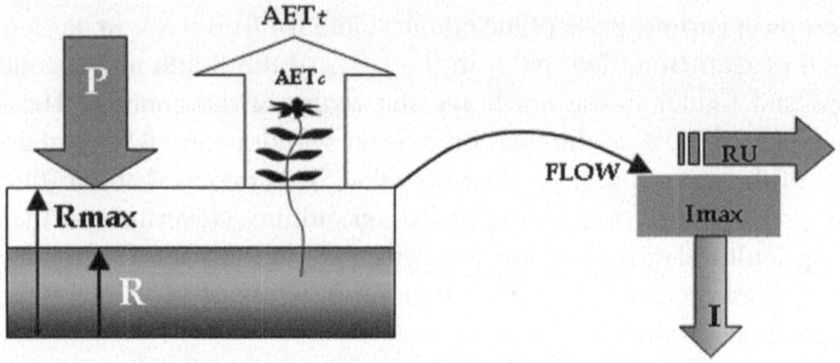

Fig. 7.1 Schematic layout of the Soil-Plant Water Balance Model

7.2 Green Water Footprint of Rainfed Agriculture

7.2.1 Modeling the Water-Soil-Plant System

The model principle

To represent locally and regionally the water balance of the soil-plant system, we use a conceptual reservoir model, a production function type, representing the superficial soil functioning with a daily time step. The model (Box 7.1) includes: (i) two entries: daily rainfall (P) measured in a representative station of the area and the daily Potential Evapotranspiration (PET); (ii) parameters defining the plant (defined by the crop coefficient): state of development; ability to draw water from soil; (iii) parameters defining the soil: Soil Useful Reserve (Rmax); deep infiltration capacity Imax; (iv) a production function that establishes the soil water balance; (v) five outputs: total actual ET (AETt) of the soil-plant system, sum of the bare soil evaporation and plant ET, state of the soil reserve R, runoff RU, deep infiltration I, Actual crop Evapotranspiration (AETc).

The production function divides rainfall at the surface between infiltration, runoff, direct evaporation and storage in the soil (Fig. 7.1). The plant draws water from the soil reservoir, according to its development schedule and its possible state of water stress function of the soil moisture degree. The system is described by a series of balance equations that determine, at the end of each time step, the state of each of the system outputs (see Box 7.1).

Reference area, inputs and outputs

We chose to work by governorate, the Tunisian administrative divisions for which there are relevant information. The model that works autonomously for each governorate admits as input variable the daily rainfall observed in one representative station[1]. The year 2004–2005, which general rainfall is close to that of an average

[1]Daily rainfall series from DGRE databases, kindly communicated to the author by H. L. Frigui, then Director of the hydrology department, September, 2012.

hydrological year, was selected to conduct this exercise. The monthly PET in governorate chief towns is that of Penman-Monteith published by Henia et al. (2008). The daily PET is obtained by dividing the monthly value by the number of days of the month. The chosen reference area for implementing the model covers the three governorates of Beja, Siliana and Le Kef situated in Northwest Tunisia (see Fig. 3.2, Chap. 3), whose total production covers almost 50% of the national cereals production, and where supplemental irrigation constitutes a negligible source of water. As reference output value for the model validation, we have only the order of magnitude of cereals Water footprint, for which many values are published in the literature; the most frequent values vary between 1.5 and 2 m^3/kg of cereals.

Box 7.1: The Soil—Plant Water Balance Model

The actual evapotranspiration of a given crop AET*c* depends on: (i) the climate data and the potential evapotranspiration PET; (ii) the crop characteristics; (iii) the soil structure and water availability in the soil. At a daily scale for a given day i, the total actual evapotranspiration $AETt_i$ equal to Min [PET_i; R_{i-1} +P_i], results from the water balance of the soil reservoir. The actual crop evapotranspiration is expressed by the equation: $AETc_i$ = Min [$AETt_i$; kc_i * ks_i * PET_i]; where kc is the crop coefficient of the plant; ks is a stress factor reflecting the difficulty that the plant has in water stress condition and PET refers to potential evapotranspiration. The value of the crop coefficient depends on the type of crop and its stage of development.

The FAO Guide (Allen et al. 1998) provides the evolution of kc for different cultures and contexts. As for the coefficient of stress ks, its expression is of the form ks_i = Min [1; (R_i/(Θ * Rmax))]; R_i being the level of the total soil water reserves in the day i, determined by the production function, iteratively with $AETt_i$ and the Useful Reserve of the soil Rmax. θ is the fraction of Rmax that the plant can extract from the ground without suffering water stress. If we admit θ = 0.5, ks is at its maximum, equal to 1, when the soil moisture reaches the Easily Usable Reserve RFU (RFU = 0.5 * Rmax). Below, ks < 1 and the plant is suffering water stress. The model parameters that can be adjusted by successive approximations are: (i) the water reserve of the soil whose achievement triggers infiltration, comparable to field capacity; (ii) infiltration capacity Imax, beyond which there is runoff.

Implementation and results achieved

For each of the three selected regions-governorates, the Table 7.1 shows all the data and results obtained at the model output, where we observe that the GWF values provided by the model range from 1500 to 2000 m^3 of water per ton of production expressed as dry matter, of cereals, which seems consistent with the values that can generally be found in the literature. Based on the likely assumption of a uniform

Table 7.1 The soil-plant water balance model, data and results

Governorate	Beja	Le Kef	Siliana
Data of the governorate			
Cereals area 1000 ha[a]	137	208	158
Average yield ton/year	2.49	1.33	1.51
Cereals production 1000 ton/year	341	277	239
Model parameters and input			
Rmax mm	100	100	100
Imax mm/day	10	10	10
Rainfall mm/year[b]	676	410	535
Model results			
Total actual evapotranspiration (AETt) mm/year	660	410	535
Crop actual evapotranspiration (AETc) mm/year	472	230	290
Productive green water m^3/year	6.5E+08	4.8E+08	4.6E+08
Green water footprint (GWF) m^3/ton	1896	1729	1921

[a]Average 2002–2007; [b]Year 2004–2005

allowance of fertilizers in governorates, these preliminary results verify and confirm that yields are directly related to rainfall input.

7.2.2 *Modeling the Rainfed Crop Production Water Footprint*

The concept of GW suggests that under given conditions, the variation of crops yields should be proportional to the variation of water amount applied to the field during the growing period. The onsite work of Rezgui et al. (2000, 2005) have been devoted to the study of the variability of the wheat production cultivated in five different bioclimatic sites in Tunisia under rainfed and irrigated conditions. The tests conducted do indeed indicate that the grains yields are linearly correlated with the net water consumption measured on the field (Rezgui et al. 2005), and suggest the existence of a minimum water consumption ensuring grain production whose value is around 210 mm/year. Presumably this threshold, which includes rainfall during the dry season, may vary by region. One can, for simplicity, admit that for a given region the threshold height is independent of the hydrological year; it is noticed h_S.

These empirical results suggest that it is possible to calculate the productive GW volume by performing a linear regression of grain production by annual rainfall accumulation during the production cycle in a given region. Once validated by statistical analysis of rainfall and crop yield data on cultivated area, the linear regression model can be used to evaluate the WF of agricultural production and to characterize its variation (Chahed et al. 2011). When the hydrological year, which begins in September, recovers all the production cycle, as for the cereals crops, one may consider that the rainfall accumulation will correspond to the annual rainfall accumulation during the hydrological year. This is not the case for the olive trees, for example, whose production cycle covers two successive hydrological years: part of the current hydrological year, considered as the production year, and part of the preceding year. In order to consider the biannual alternate-bearing cycle of olive trees productions, one may consider that the two successive years are involved in the production of the year (n) with two respective weights α for the year (n) et β for the year $(n - 1)$ so that $\alpha + \beta = 1$.

Box 7.2: A Linear Regression Model to Estimate the Green Waterfootprint of Rainfed Crop Production

The amount of green water into play in the rainfed production during the hydrological year (n) in a given area, expressed in height of water (volume per unit area) and noticed $h_G(n)$ represents the portion of rainfall effectively used by plants or efficient rainfall: $h_G(n) = (h_R(n) - h_S)$ where $h_R(n)$ is the rainfall of the hydrological year (n) and h_S is the height limit below which the yield $Y_G(n)$ is zero. It follows that $Y_G(n)$ is directly proportional to $h_G(n)$. By choosing to express the yield $Y_G(n)$ in Quintals per hectare and $h_G(n)$ in mm, we can write: $h_G(n) = h_R(n) - h_S = 10GWF_P \times Y_G(n)$ where GWF_P is the specific green water footprint in m^3 per kg of agricultural product (p) in the region concerned.

The same analysis extended to all regions of the country leads to similar results. The Green Water volume nationwide naturally expresses the sum of regional contributions. This is equivalent to applying a spatial averaging operator $(\overline{\ })$ by weighting the quantities by the part of cultivated area in each region. Moreover, if one assumes the part of cultivated area in each region is equal to inter-annual mean value one can determine the average amount of green water volume at the national scale. The corresponding spatio-temporal operator will be noted by $\langle\overline{\ }\rangle$. The green water volume (m^3) of rainfed agriculture, which occupies an average area $\langle A\rangle$in hectars, is obtained using the following two formulations:

$$\langle\overline{GW_1}\rangle = 10\left(\langle\overline{h_R}\rangle - \overline{h_S}\right)\langle A\rangle \tag{7.1}$$

$$\langle\overline{GW_2}\rangle = 100\overline{GWF_P}\langle\overline{Y_G}\rangle\langle A\rangle \tag{7.2}$$

Figure 7.2 indicates a significant change in cereal yields over the past four decades with a small stabilization in the last period 2001–2010. The cereal farming was intensified by a set of measures applied by extension and supervision of farmers: use of certified seed, fertilizer, chemical weed control, preventive treatments. However the linear regression model does not incorporate yields improvements and its adjustment should be carried out over relatively short periods of time to consider that the change in average crop yields remains relatively low.

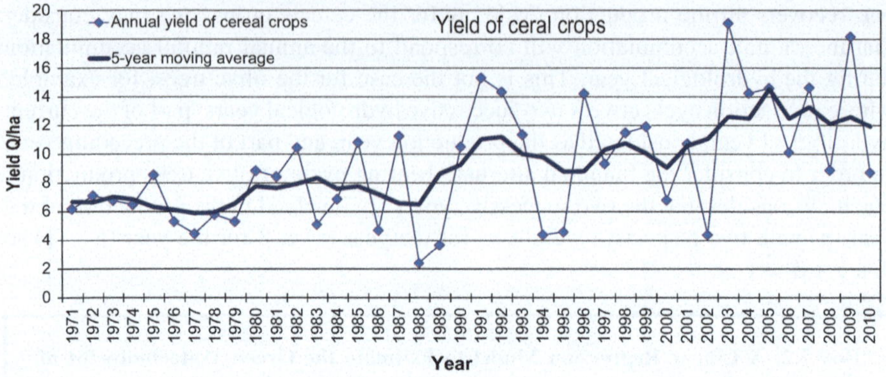

Fig. 7.2 Change in cereal crops yield, in Q/ha, from 1971 to 2010

7.2.3 Water Footprint of Cereal Crops Production

Regional scale analysis

Based on regional agricultural statistics, it is proposed to check whether the formulation of GWF has a statistically acceptable signification for cereal crops at regional and national scales. In the national formulation, quantities are weighted by regional proportions of cultivated areas. As in the previous section, we identify the region to administrative boundaries of the governorate. This administrative division also corresponds to the regionalization adopted in agricultural statistics published by the Ministry of Agriculture.

In a first step, the model is applied to cereal crops in the major producing governorates (Beja, Siliana and Le Kef) adopting hydrological and production data over a period of seven years (2001–2007) around the period for which the soil-plant system water balance model has been implemented.

Figure 7.3 shows the evolution of all cereal crops yields as a function of annual rainfall depth in these regions. The model is based on the assumption of uniform fertilizer application in the three regions. The resulting curves show that one can find a linear relationship between rainfall and yields of cereal crops. The linear regressions, however with uneven qualities, make it possible to determine the model parameters. The x-intercept of the regression line provides the threshold and the inverse of its slope is used to determine the specific GWF (GWF_P) of cereals productions.

These parameters are presented in Table 7.2 where it is possible to compare the values of specific WF of cereals production in different regions with the parameters obtained from modeling water balance of the soil-plant system. It appears that the linear regression model provides for the different regions, orders of magnitude comparable to values obtained from the soil-plant water balance model.

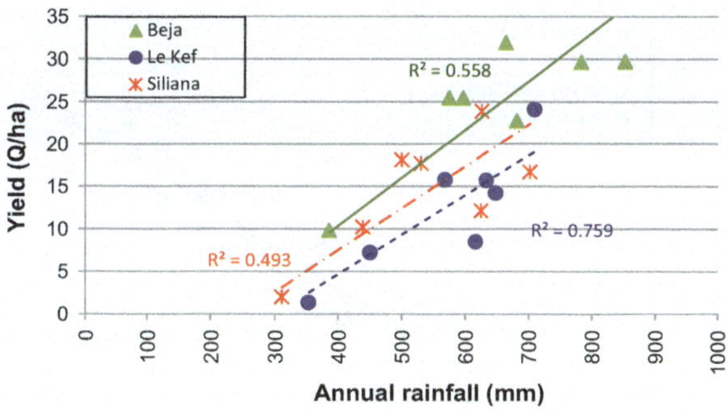

Fig. 7.3 Cereal crops yields in major producing governorates as function of annual rainfall (2001–2007)

Table 7.2 Parameters of linear regressions in major cereals producing governorates (2001–2007)

Governorates	Annual rainfall ⟨h_R⟩ (mm)	Threshold ⟨h_S⟩ (mm)[c]	Specific GWF[a] GWF_P (m^3/kg)[d]	Specific GWF[b] GWF_P (m^3/kg)
Beja	658	222	1.8	1.9
Le Kef	589	320	2.0	1.7
Siliana	550	249	2.0	1.9

[a]Linear regression model. [b]Soil-plant water balance model. [c]x-intercept of the regression line. [d]Calculated as the inverse of the regression line slope

Nationwide scale analysis

The analysis is extended to national cereal crops by weighting local parameters by cultivated areas, in accordance with the model formulation and by exploiting the national cereal production statistical data.

As mentioned above, the linear regression model does not incorporate yields improvements. It is therefore limited to the decade 2001–2010 assuming that average crop yields have no significantly changed during this period. It is observed that the nationwide correlations (Fig. 7.4) are of much better quality than those at the regional level (Fig. 7.3); indeed, weighting rainfall by regional surfaces results in a form of smoothing which provides best correlations. Nevertheless, and given the adopted assumptions and resulting approximations, the results seem to constitute reasonable and acceptable estimates, compatible with the scale and the general character of the regression model.

According to the model assumptions, cereal crops yields are expressed in terms of useful weighted rainfall according linear laws with statistically significant correlations. The correlation parameters were determined for the major cereal crops (durum wheat, bread wheat, barley) and for all cereals. These parameters are used to calculate the useful rainfall depth related to GW, or the productive GW, for the considered crops. Applied to the cultivated areas, these rainfall depths are used to

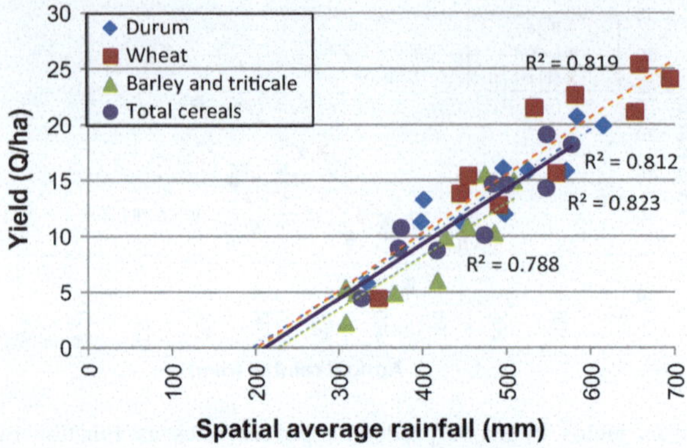

Fig. 7.4 Nationwide cereal crops yields as a function of weighted rainfall, (Period, 2001–2010)

Table 7.3 Nationwide linear regressions parameters (period, 2001–2010)

Culture	Weighted rainfal $\langle h_R \rangle$ (mm)	Water threshold[a] $\overline{h_S}$ (mm)	Specific $\frac{GWF^b}{GWF_P}$ (m^3/kg)	Green water height $\langle h_G \rangle$ (mm)	Sawn area (10^3 ha)	Green water (km^3)
Durum	487	198	2.1	289	735	2.1
Wheat	542	204	2.0	338	132	0.4
Barley	407	228	2.1	179	526	1.0
Total					1392	3.5
Total cereals	462	210	2.0	252	1392	3.5

[a]X-intercept of the regression line. [b]Calculated as the inverse of the slope of the regression line

calculate the GW average volumes of cereal crops. All of the results are summarized in the Table 7.3.

Analysis of the GWF of national cereal productions over the period 2001–2010 indicates that the mean annual rainfall weighted by the sown area in the different regions has a value of about $\langle h_R \rangle = 462$ mm. As the rainfall threshold is about $\overline{h_S} = 210$ mm, the GWF of rainfed cereal crops production represents a useful rainfall depth of $\langle h_G \rangle = 252$ mm corresponding to a specific volume of 2520 m^3/ha. Reported to the national average cereal yield, this volume corresponds to a GW specific volume of around 2 m^3/kg. It should be noted that the correlations obtained for each cereal crops provide in total, the same GWF calculated for all cereals production. These results confirm and validate those obtained by the soil-plant water balance model.

Discussion of the Results

BWF and GWF represent the consumptive use components of each of the two kinds of water, Blue and Green. By determining a direct relationship between crop production and precipitation, the linear regression model determines the productive component of GW (GWF) involved in rainfed agriculture production. The model parameters are directly related to water use efficiency: the rainfall threshold $\overline{h_S}$ might be said to represent the non-productive component of GW (mainly evaporation); the useful rainfall depth (the difference between rainfall $\overline{h_R}$ and the threshold $\overline{h_S}$) is thus a direct indicator of the GW consumption. This productive part is converted into crop production at the rate $\overline{GWF_P}$ representing the specific amount of consumptive water per kg of crop production which will thus reflect the water use efficiency in relation with agricultural practices (plant varieties, fertilizers…). Advancement of GW management has to be sought in an integrated approach of the water-land system which aims at reducing the non-consumptive part of GW ($\overline{h_S}$ parameter in the model) and the specific amount of consumptive water per kg of crop production ($\overline{GWF_P}$ parameter in the model).

In general, available data on water resources only include BW and information on GW is scarce. Production of reliable data about GW represents a major challenge. According to the linear regression model, the WF of rainfed crop production concerns the cultivated area, apart from the fact that cultivated crops give rise to harvests or not. Indeed, some databases refer to harvested areas, especially for field crops (cereal and forage crops). For example FAOSTAT data bases related to cereals consider only harvested area for dry grain production, and cereal crops harvested for hay or harvested green for food, feed or silage or used for grazing are therefore excluded. This type of data does do not take into account the areas sown that have not produced enough crops to be harvested. Accordingly, it does not take into account yield losses associated with drought episodes.

Over the period 1971–2010, Fig. 7.5 compares the sown areas (Ministry of agriculture data base) with harvested areas (FAOSTAT), between which there is a steady and constant gap of nearly 15%, about 200,000 ha. Since production is the same in both data sources, this results in significant differences in the calculated yields. In order for the linear regression model to be compatible with the WF concept, its adjustment to the different scales (regional and national) must concern the cultivated areas as a whole, thus integrating all the GW resources involved in agricultural production and enabling to analyze their annual variability.

7.2.4 Green Water Footprint of Olive Oil Production

The total area of olive plantations amounts to 1.666 million ha, almost all of which (97%) is rainfed (Karray et al. 2009). Exclusively olive oil planted acreage is estimated at 1.375 million hectares cultivated mainly under rainfed conditions. The olive crop cycle spans two hydrological years. From March to June, the need for water and nutrients of the tree are the most intense. During the summer, the olive tree goes through a period of development and growth of the olives, which ends in the autumn by a maturation phase of the fruit and its oil enrichment. The olive harvest lies between late autumn and early winter.

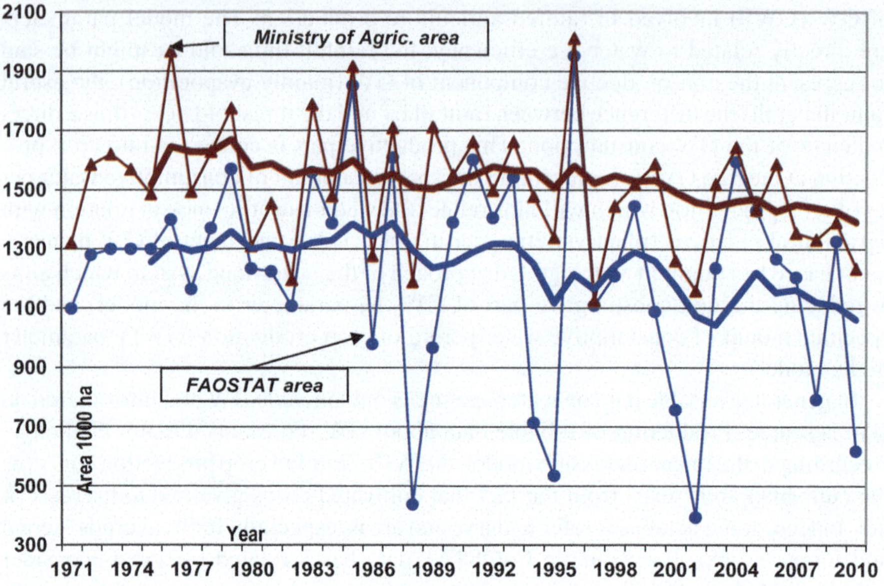

Fig. 7.5 Areas of cereals, sown, (Ministry of Agriculture 2012) and harvested (FAO 2017), in 1000 ha, and trend curves (5 years moving average)

The olive crop cycle therefore covers two hydrological years: the year of harvest and the year preceding it. To take into account the biannual alternate-bearing cycle of olive trees productions, we consider that two successive years are involved in the production of the year (n). This defines the weighted rainfall depth of one year covering the cycle of the olive tree production by $h_W(n) = \alpha h_R(n) + (1 - \alpha)h_R(n - 1)$ where α is the weight related to the year (n) and $(1 - \alpha)$ that related to the year $(n - 1)$. It is admitted that in each region, the production is proportional to the rainfall depth. The average depth across the country is calculated by weighting the rainfalls according to the regional plantations surfaces.

The implementation of the model to the olive oil farming covers the period 1996–2004. The adjustment of the coefficient α shows that the better correlation is obtained for the value $\alpha = 0.25$, suggesting that in average, the hydrological year of olive production contributes to about one quarter in feeding olive trees while the preceding year is for almost three quarters. The results for the evolution of the yields per hectare of the national olive production as a function of the weighted rainfall by regional olive plantations areas on the period 1996–2004 are presented on Fig. 7.6.

The results (Table 7.4) indicate that the threshold stands at $\overline{h_S} = 186$ mm and the depth of annual precipitation weighted by regional areas of olive plantations is $\langle \overline{h_W} \rangle = 294$ mm. The average volume of GW associated with the production of oil olive is then equal to $\langle \overline{h_G} \rangle = 108$ mm, which represents a specific volume of 1080 m^3/ha. Applied to the average full olive-growing area, estimated at about 1.375 million hectares, the volume of GW involved in the production of olive oil, averaged

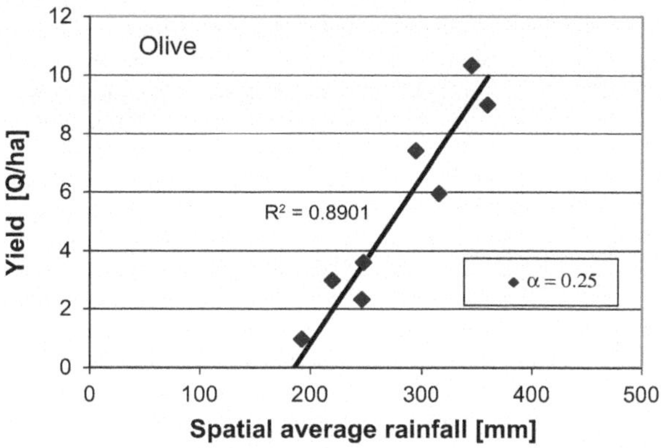

Fig. 7.6 Production yield of olive as a function of to the weighted rainfall depth ($\alpha = 0.25$), period (1996–2004)

Table 7.4 Parameters of the correlations between oil olive plantations yields and the weighted rainfall depths ($\alpha = 0.25$)

Crop types	Weighted rainfal $\langle h_W \rangle$ (mm)	Water threshold $\overline{h_S}$ (mm)	Specific GWF $\overline{GWF_P}$(m^3/kg)	Green water height $\langle h_G \rangle$ (mm)	Plantation area (10^3 ha)	Green water (km^3/year)
Olive trees	294	186	1.75	108	1375	1.5

over the period 1996–2004, amounts to 1.5 km^3/year. The correlation also indicates that the specific GW volume amounts to 1.75 m^3/kg of olive.

Recall that the Tunisian olive groves densities (olive trees per hectare) vary widely: relatively high in the northern humid regions where the trees density reaches almost 100 trees/ha, it decreases significantly in the central regions of the country (about 50 trees/ha in the Sahel region) and falls below 20 plants/ha in the Sfax region. The density of olive plantations has in some way coped with the available water resources in relation to the rainfall patterns: rainfall decreases in amount and regularity from north to south across the country.

Applying the total GW volume to the total olive plantations estimated at 65 million trees, the average specific volume is estimated to be around 22 m^3/year/tree. There is no specific literature that would allow a systematic validation, but some comparisons can be made with Spanish data. Salmoral et al. (2010) estimated the equivalent GW of Spanish olive rainfed plantations to 1971 m^3/year/ha which represents a useful rainfall depth of 197 mm, nearly two times higher than the average value found for Tunisian plantations. As the Spanish olive groves have a relatively high density planting, typically 100 trees/ha, the GW specific volume stands at nearly 20 m^3/year/tree, very close to the values calculated for Tunisian olive trees.

Table 7.5 Green water potential of rainfed agriculture in Tunisia

Crops types	Weighted rainfall $\langle \overline{h_W} \rangle$ (mm)	Water threshold $\langle \overline{h_S} \rangle$ (mm)	Green water depth $\langle \overline{h_G} \rangle$ (mm)	Maximun rainfed area (10^3 ha)	Green water potential (km^3/year)
Cereals	462	210	252	2023	5.1
Olive trees plantations	294	186	108	1666	1.8
Other rainfed crops			187	834	1.6
Rainfed agriculture	Total			4523	8.5

7.2.5 The Green Water Potential of Rainfed Crops

Cereal crops are widely practiced in the humid northern regions, while the olive groves are located in the central regions where rainfall is much lower. It follows that rainfall mobilized by olive groves per unit area is much lower than the volume used per unit area of cereal crops.

If we apply the useful rainfall depth (GW) associated to cereal crops (252 mm) to the total cultivated area during 1996 (2.023 million ha), which corresponds to the maximum area sown between 1984 and 2010, we may estimate the potential of the Tunisian cereals agriculture at about 5.1 km^3/year. Likewise, if we apply the useful rainfall depth (GW) related to olive groves (108 mm) to all Tunisian olive groves whose area totals to 1.666 million hectares, one can estimate the potential of GW associated with olive plantations at about 1.8 km^3/year.

On average, the potential associated with these two main cultures is nearly 6.9 km^3/year collected on a total area of 3.7 million hectares, which represents an average specific volume of 1870 m^3/ha. Assuming that specific volume applies to all cultivated areas under rainfed conditions whose potential is nearly 4.5 million hectares (maximum area of rainfed cultures between 1984 and 2010), one can estimate the potential of GW of rainfed agriculture to be about 8.5 km^3/year. Note that this resource drawn by rainfed agriculture from soil water is four times greater than BW resource allocated to irrigation. The most significant results related to GW potential of rainfed agriculture are summarized in Table 7.5.

7.2.6 Variability of Green Water Resources

The calculation of the useful rainfall depth (GW) was used to assess the water potential of rainfed agriculture using the Eq. 7.1 (Box 7.2). The parameters obtained from the correlation between precipitation and production of rainfed crops allow, using the Eq. 7.2 (Box 7.2), to provide a new assessment of the volume of GW nationwide. The

Table 7.6 Comparison of assessment methods of green water footprint to the output of the major rainfed crops

Crop types	Average cultivated area (10^3 ha)	Average production (10^3 tons)	Green water height $\langle h_G \rangle$(mm)	Specific GWF $\overline{GWF_P}$ (m^3/kg)	Green water (km^3)	
					Equation 7.1	Equation 7.2
Total cereals	1392	1762	252	2.0	3.6	3.5
Olive	1375	831	108	1.75	1.5	1.5

Table 7.6 compares GW volumes calculated by the two formulations, respectively for cereal crops (2001–2010) and the olive sector (1996–2004). This result indicates that both equations (Eqs. 7.1, 7.2) provide completely concordant evaluations of the average GW related to rainfed crops production.

It should be noted that the comparison made in the Table 7.6 does not constitute a validation of the model because the parameters of the two formulations (Eqs. 7.1, 7.2) were estimated using the same statistical data. This simply confirms that the production of rainfed crops and average GWF evolve in a homothetic manner. The consistency check of the model is however very useful to study the variability of the GW resources. Indeed, by relying on statistics relating to the production of rainfed agriculture, we can evaluate the GWF not only in average, but also in statistical quantities characterizing its fluctuation, so it becomes possible to associate occurrence frequencies to extreme values (minimum and maximum).

Table 7.7 shows that the annual variability of GWF is significant. For all cereal crops, the standard deviation of GW volumes has an order of magnitude of 40% of its average value (around 37% for wheat and 60% for barley), while deviations of the weighted rainfall is much lower. They represent less than 20% of the average for all cereal crops (19% for wheat and 20% for barley). Regarding olive groves, the standard deviation of the GW volumes amounts to almost, 50% of the average value, while the weighted rainfall depth has a standard deviation of about 25% of its average value. By characterizing the variability of the GWF of rainfed crops productions, it becomes possible to evaluate extreme values for different cultures and for their productions GWF, and to associate to them occurrence frequencies. The results are shown in Table 7.8.

In general, the variability of the GW resources associated with rainfed crops production is roughly two times higher than that of the rainfall. It is interesting to note that the variability in production of the cereal crops does not only depend on rainfall variability, it also depends on the variability of the sown areas. The later are also correlated with rainfall and its variability has obvious implications on the assessment of GWF of rainfed crops. It also appears that the variability in production of barley is higher than that for other cereals due to a higher variability of rainfall in areas where barley is preferably cultivated. It's the same for the olive-growing regions of the country where the fluctuations of GWF are higher than those for cereals (see

Table 7.7 Average values and standard deviations of weighted rainfall, production and green water footprint of the major rainfed crops

	Weighted rainfall $\langle \overline{hw} \rangle$ (mm)		Production (10^3 ton/year)		Green water (km^3/year)	
	Average	Standard deviation	Average	Standard deviation	Average	Standard deviation
Total cereals	462	86	1762	742	3.6	1.5
Olive	294	72	831	389	1.5	0.7

Table 7.8 Variability of green water resources

Crop types	Green water average volume (km^3/year)	Standard deviation (km^3/year)	Green water (km^3/year)			
			Frequency 20% Max–Min	Frequency 10% Max–Min	Frequency 5% Max–Min	Frequency 2% Max–Min
Total cereals	3.6	1.5	4.7–2.3	5.5–1.7	6.1–1.1	6.7–0.5
Olive	1.5	0.7	2.1–0.8	2.3–0.6	2.6–0.4	2.9–0.1
Rainfed agriculture (potential 8.5 km^3/year)			11.6–5.4	13.2–3.8	14.5–2.4	16.1–0.9

Table 7.8). The variability of olive production is amplified by the biennial cycle of the olive trees which causes considerable fluctuations in their production.

7.3 Assessing Water Footprint of Crops Production

We propose in this section to assess the WF of all national agricultural production. Based on statistical data on crop production (rainfed and irrigated), we try to assess their WF. In the first part, we develop a model of the WF of irrigated crop production based on that developed for rainfed agriculture. The assessment methodologies are adapted according to the available data and applied to the various elements of the national agricultural production.

7.3.1 *Water Footprint of Irrigated Crops Production*

Modeling the water footprint of irrigated crops production

The same approach developed for rainfed agriculture leads to a similar formulation of the linear regression model for irrigated crops. This model expresses the yield of irrigated agriculture $Y_I(n)$ based on rainwater resources and artificial inputs of irrigation water. The contribution of direct rainfall is highly variable and depends on climatic conditions; it is almost negligible in desert zones. The distinction between GW and BW in agricultural production makes it possible clarify the WF of agriculture on ecosystems and natural resources.

As for rainfed crops, WF production of irrigated crops EI(n) (in m^3) is proportional to the cultivated area $A_I(n)$ (in ha) so that it is possible to express the water depth effectively used by crops. This depth includes (i) the depth $h_I(n)$ (in mm) corresponding to the water depth effectively used by irrigation, after deduction of losses on the parcel, of direct evaporation and of deep infiltration; and (ii) the useful rainfall depth $h_G(n)$ that corresponds to the GW input which as for rainfed crops, is calculated as the difference between the rainfall over the period covering the whole crop cycle $h_W(n)$ and the threshold depth from wherein there is production ($h_G(n) = h_W(n) - h_S$).

The linear regression model formulated for irrigated farming in a given region writes: $h_I(n)+h_G(n) = 10WF_P \times Y_I(n)$. Where WF_P is the specific WF in m^3 per kg of crop product. The model equations are used to distinguish between the contributions of GW and BW to the WF of crop production in order to clarify the various balance sheet items. As for rainfed agriculture, the weighted rainfall by cultivated areas can be calculated for each rainfed and irrigated agriculture, using the distribution of cultivated areas and regional hydrologic data. Field experiences for cereal crops and analysis developed in the previous section for rainfed crops suggest that the threshold level is established at a depth of about $\overline{h_S} = 200$ mm. We admit that the same holds true for other cultures.

Evaluation of the effective water depth of irrigation

It is possible to estimate the effective water depth related to irrigation for different crops based on crops water needs by region, applied to the spatial distribution of irrigated areas of different cultures through the whole of the country. The Ministry of Agriculture (DGGREE) defines these regionalized water needs, for various crops (MARH-DGGREE 1995). These data indicate that in general the needs of crops, forage crops, legumes and winter vegetable crops are pretty moderate about 2000–3000 m^3/year/ha while the water needs of summer vegetable crops and tree crops are relatively high (5000–6000 m^3/year/ha).

The highest needs are those of the date palm in the oases with volumes of around 15,000 m^3/ha. It should nevertheless be noted that because of losses in the supply and distribution networks, the volumes allocated are greater than the water volume delivered to the parcel. On the other hand, the water resources delivered to the parcel are only partially used by the plants and the efficiency of use of water depends on the mode of irrigation. If one accepts that the performance of hydraulic networks is

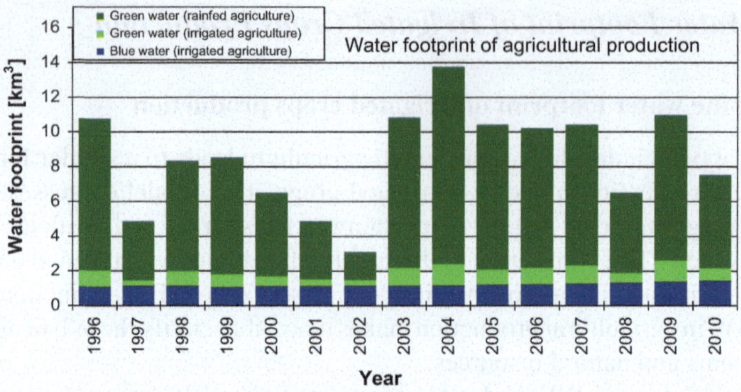

Fig. 7.7 Evolution of blue water footprint and green water footprint of crops production

80% on average and that the effectiveness of the use of water at the plot amounts to 70–80%, the overall efficiency of the irrigation would be on a national average of around 60%.

7.3.2 *Water Balance of Crops Production*

The methods for evaluating the WF of rainfed and irrigated agriculture will be applied to estimate the WF of the different cultures. It relies for this on national agricultural statistics during the period 1996–2010. This makes it possible to clarify the contributions of rainfed and irrigated agriculture to WF of crop production and to analyze its trends over fifteen years.

The WF assessments of rainfed and irrigated crop productions allow the estimation of the WF of all crop production. Figure 7.7 shows the evolution of the WF of crop productions where can be distinguished the contributions of BW and GW for irrigated and rainfed crops. There appears (see Table 7.9 too) that the WF of all crops production represents an average amount of water resources of over 9 km^3 in the late 2000s.

This amount is however subjected to considerable variations mainly due to the variability of rainfed production in relation to rainfall variability. The WF of rainfed production has increased from less than 2 km^3 in 2002 to over 11 km^3 in 2004. At the opposite, variability of WF of irrigated crops production is much low. During the fifteen year between 1996 and 2010, the WF of the irrigated agriculture production has been limited in the range of 1.5–2.5 km^3, highlighting the role of irrigation in the regulation of the agricultural production. The amount of BW used in irrigation remains regular and gradually increases from 1.1 km^3 in 1996 to over 1.5 km^3 in 2010; this growth should be attributed to the irrigated agriculture development.

Table 7.9 Average water footprint of crop production, (2006–2010)

	Water footprint of crops production (million m^3—average 2006–2010)				
	Rainfed agriculture	Irrigated agriculture		Total green water	Total
		Green water	Blue water		
Cereal crops	3220	169	90	3389	3479
Forage crops	942	175	92	1117	1209
Leguminous crops	307	10	4	317	321
Vegetable crops	36	285	432	321	753
Olive tree	1619	50	74	1668	1742
Date palm	0	0	338	0	338
Fruit tree	703	180	331	883	1214
Industrial crops	59	17	12	75	87
Total agricol production	6885	885	1372	7770	9142

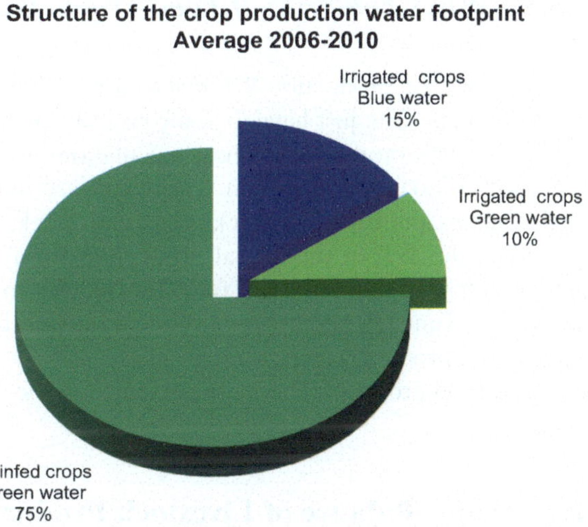

Fig. 7.8 Structure of water footprint of crop productions

Figure 7.8 shows the structure of the average WF of agricultural production during the period 2006–2010. The contribution of BW constitutes approximately 15% of the WF of the agricultural production. GW thus amounts to 85% of water resources involved in agricultural production. This resource, which is the WF production of rainfed crops (75% of water resources) and partly that of irrigated crops (10% of water resources) is subject to considerable variations as it depends directly on rainwater resources.

On average, the WF of irrigated agriculture is finally only a quarter of water resources involved in the total agricultural production. In spite of its essential contribution in different sectors, the irrigated agriculture is too limited to be able to stabilize the entire agricultural production. In the WF, the contribution of rainfed agriculture represents on average three quarters of agricultural production. High variable, its production in the different sectors is not able to respond to the food demand and Tunisia has an inbuilt need for additional contributions to fill the gap between the local production and the food demand, especially during periods of drought or low rainfall.

Table 7.9 highlights the structure of the mean water balance of agricultural production during the years 2006–2010. It appears that, in terms of WF, cereals, forages and olive productions represent 85% of the GW resources involved in rainfed agriculture. The WF of the production of vegetables and crop trees, including date palm, represents 70% of the WF related to the production of the entire irrigated area and employs 80% of BW resources involved in irrigation. The fruit tree growing is the largest beneficiary of the water resources of the country, it accounts for 50% of the resources in BW half of which is allocated to the cultivation of date palm.

7.3.3 *Water Footprint of Food Products*

Analysis of crop production WF makes it possible to determine WF of the various plant products. One just have to confront water volumes consumed by the different cultures to their productions. These calculations are used to specify contributions of BW and GW to crop production. The results are shown in Table 7.10 and compared to the values published by Mekonnen and Hoekstra (2010). These authors have quantified the WF of the global production for 126 crops using a water balance model with spatiotemporal discretization similar to that used above for calculating the WF of Tunisian cereal crops production (see § 7.1). The WF of the different agricultural products can be used as an assessment basis for estimating specific WF of major food products. These are presented in the Table 7.11.

7.4 Water Balance of Livestock Productions

7.4.1 *Water Footprint of Livestock Productions*

The WF of animal products can be determined from the WF of products used for animal feeding. In Tunisia, a significant proportion of fodder and grain used for animal feeding are imported and another significant part comes from extensive grazing in relation to pastoral activity and for which there is not enough data to assess its contribution. It is therefore proposed to draw on comprehensive methods of evaluation used

Table 7.10 Specific water footprint of major agricultural products in Tunisia according to present calculations (2006–2010) and in the world (last column)

Average 2006–2010	Water footprint (10^6 m^3)	Production (10^3 ton)	Specific water footprint (m^3/kg)			Specific WF[a] (m^3/kg)
			Blue water	Green water	Total	
Cereal crops						
Durum wheat	2054	991	0.050	2.023	2.073	
Soft wheat	424	226	0.041	1.835	1.876	1.639
Barley and triticale	1001	462	0.070	2.095	2.165	1.292
Forage crops						
Oat	582	723	0.061	0.744	0.805	1.560
Legumes						
Bean and horse bean	202	57	0.039	3.527	3.566	1.522
Peas and chickpeas	82	20	0.046	4.115	4.161	3.196
Vegetable crops						
Potato	89	357	0.113	0.135	0.248	0.224
Tomato	155	1103	0.093	0.047	0.140	0.171
Pimento	118	283	0.275	0.141	0.416	0.282
Onion	47	358	0.050	0.079	0.129	0.220
Artichoke	11	17	0.313	0.337	0.651	0.720
Watermelon and melon	99	493	0.118	0.083	0.201	0.172
Other vegetable crops	234	347	0.377	0.300	0.677	
Tree crops						
Olive	1742	900	0.082	1.853	1.936	2.969
Dry almond	254	56	0.928	3.645	4.573	6.540
Dates	338	147	2.293	0.001	2.294	2.180
Citrus	129	283	0.261	0.196	0.456	0.511
Grapes	141	122	0.615	0.532	1.148	0.522
Other tree crops	690	428	0.306	1.307	1.613	
Industrial crops						
Sugar beet[b]	7	55	0.060	0.067	0.127	0.108
Other industial crops	87	57	0.203	1.324	1.527	

[a]Global average according to Mekonnen and Hoekstra (2010)
[b]Average 1996–2000

Table 7.11 Water footprint of agrifood products

Agrifood products	Specific water footprint (m^3/kg)		
	Blue water	Green water	Total
Olive oil	0.410	9.270	9.680
Sugar	1.198	1.344	2.542
Canned tomatoes	0.558	0.282	0.84
Fruit juice	0.306	1.307	1.613
Canned fruit	0.612	2.614	3.226
Harissa	0.605	0.310	0.915
Wine	0.861	0.745	1.607
Beer	0.023	0.698	0.722

Table 7.12 Water footprint of livestock products

Product	Conversion factor kg of cereals/kg of product	Specific water footprint m^3/kg of product
Cereals		2.0
Beef	10:1[a]	20
Sheep	6.67:1[a]	13.3
Poultry	4:1[a]	8
Eggs	2.5:1[b]	5
Milk	0.67:1[c]	1.33

[a]FAO. (1997a). [b]Maulfair (2011). [c]FAO (2009)

internationally to estimate the WF of agrifood products of animal origin. The FAO relies on the feed/meat conversion factor for estimating the WF for animal products. Considering that the WF of wheat is 1.5 m^3 per kg on average, FAO estimates the WF of beef to 15 m^3 per kg (conversion factor of 10: 1), that of poultry to 6 m^3/kg (Conversion factor 4: 1) and that of sheep to 10 m^3/kg (Conversion factor 6.67: 1). Following the approach of FAO, and using the same conversion factors, one can rely on present estimates of the WF of cereals in Tunisia, amounted to approximately 2 m^3 per kg, to estimate the animal productions. The resulting estimates are shown in the Table 7.12.

The production of agrifood products of livestock origin has continuously increased during the four last decades with a stabilizing trend during the decade 2000–2010. Tunisia has indeed reached during this latter period, self-sufficiency in animal production through a sustained policy including incentives for forage production. The specific values of WF presented in Table 7.12 have been used to estimate the WF of animal production. The corresponding calculations of WF of products of animal origin (meat, milk and eggs) are shown in Fig. 7.9.

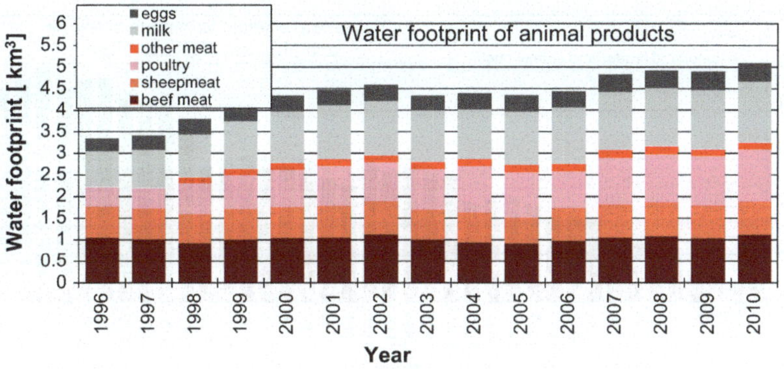

Fig. 7.9 Water footprint of food production of animal origin

7.4.2 Water Footprint Structure of Animal Production

The WF of animal production includes the WF of forages locally produced under rainfed and irrigated conditions (BW and GW), the WF of imported food products (VW) and the WF of natural pastures production. The first two items may be assessed from available statistical data on agricultural production and trade but it is difficult to assess directly the WF of livestock production from rangelands. The WF of animal production from grazing or free-range feeding could be described as GW, given that vegetation developing on rangelands and pasture is only watered by rainfall.

Rangelands and pasturages extend over large areas (4.7 million ha) but their pastoral value is generally low. The GWF of their production can be estimated by comparing the total WF of national animal production with the WF of all animal feeding. This comprises the WF of local forages and animal feed grains production as well as the VW related to animal feed imports. The difference represents the GWF of rangeland which thus appears as an adjustment variable for closing the national balance sheet of animal nutrition. However, storage of fodder and animal products is equivalent to a form of VW storage that allows inter-annual regulation by laminating the production peaks (or even of imports) in a favorable year to delay its use in an unfavorable year. Its contribution, zero in average, is not taken into account in the water balance of animal production. As a result, the closure of the balance sheet of animal nutrition only makes sense vis-à-vis its interpretation in terms of interannual average.

All local products for animal feeding, including the contribution of rangelands, represent, in terms of WF, less than 2/3 of the animal production. The sector of livestock production has an inbuilt need for an additional supply of fodder in the form of VW. The self-sufficiency in agrifood products of livestock origin displayed by the sector for over a decade is after all relative: from WF point of view, the deficit of the sector amounted to more than a third. Indeed, it appears that the performance

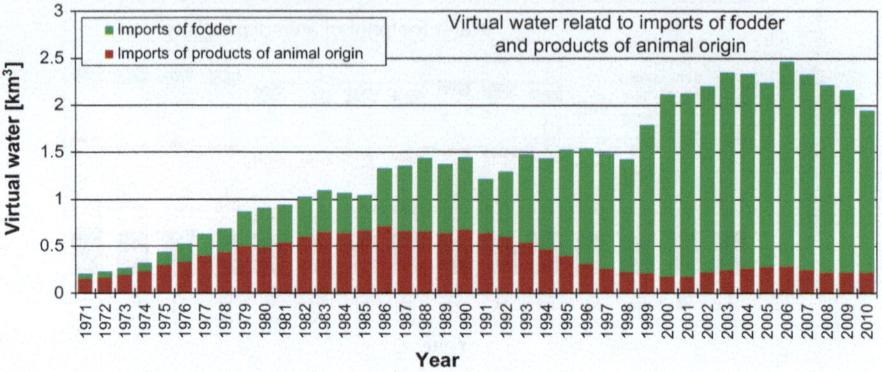

Fig. 7.10 Change in the virtual water involved in the imports of fodder and products of animal origin

of the sector has been made possible thanks to the contribution in VW associated with the fodder and cereals imports intended for animal feeding.

Figure 7.10 shows the average trend (moving average over 5 years) of VW related to the imports of agrifood products of livestock origin and the imports of fodder and cereals intended for animal feeding. It expresses, in effect, that the decrease in imports of agrifood products of livestock origin that began in the late 1980s was accompanied by a significant amplification of animal feed imports in the early 1990s. The result is, in terms of water balance, a significant increase of VW imports to support the sector of local livestock production. Within a decade these imports almost doubled going from nearly 1 km^3 in the early 1990s to about 2 km^3 in the last 2000s and since it seems that the imports are stabilizing.

Table 7.13 presents the results of calculations of the different contributions to the average WF of animal production. It specifies the contributions in WF of forage or feed cereals productions, the VW involved in the import-export balance sheet of products intended for animal feed and the WF intake of rangeland. One can distinguish between the production of irrigated crops and rainfed crops. The WF of irrigated crops production is declined in BW and GW. The contribution of irrigated sector remains low: the share of BW represents only 2–3% of the total WF of animal productions.

All of these results show that it is the GW, in all its forms (coming from crops or pasture), which provides the essential share of fodder production. Rangelands accounts for over a quarter of the water balance of animal production. This corresponds to the equivalent of almost 1.7 million tons of oats which correspond to more than twice the national average production. This significant contribution is nevertheless provided by very large areas (4.7 million ha). It is known that the livestock sector is by far the largest anthropogenic user of land and this is particularly true in arid and semi-arid regions.

Table 7.13 Average water footprint of animal productions, (2006–2010)

Products	Water footprint of animal productions (million m^3/year—average 2006–2010)					
	Blue water	Green water			Virtual water[a]	Total
		Irrigated	Rainfed	Total		
Bovine meat	24	44	365	409	325	734
Sheep meat	18	32	274	306	244	750
Poultry	25	45	374	419	333	752
Other meat	4	7	56	63	50	113
Milk	31	56	469	525	417	542
Eggs	9	17	139	156	123	179
Pasture				1348		1348
Total	111	201	1677	3226	1492	4829

[a]Virtual water represents the algebraic sum imports-exports in Water-Equivalent

7.5 Water Balance of Food Demand

7.5.1 Water Requirements for Food Demand

This section is devoted to the analysis of the water requirements for all food needs. Attempts are made to assess these needs and to analyze their change. Efforts are also made to clarify the different forms of water resources involved in crop and animal productions and to characterize their origin. This analysis is intended to draw up the water balance of food needs and to highlight key aspects of the relationship between agriculture and water resources. The analysis of the specific WF of the different agricultural speculations has led to the assessment of the water resources involved in their productions. It also clarified the origin of agricultural production (irrigated and rainfed crop culture) and the nature of water resources involved in (BW and GW).

The same approach has been used to evaluate the WF of animal productions: assessing the WF of pastures and of animal feed resulting either from local crop production or from agrifood trade. This led to an estimation of the WF of animal origin foodstuffs (meat, milk, eggs). Furthermore the VW flow involved in foodstuffs trade balance can be evaluated by applying specific VW to the food trade balance. All these elements, once assembled, will help establish the overall water balance of food demand.

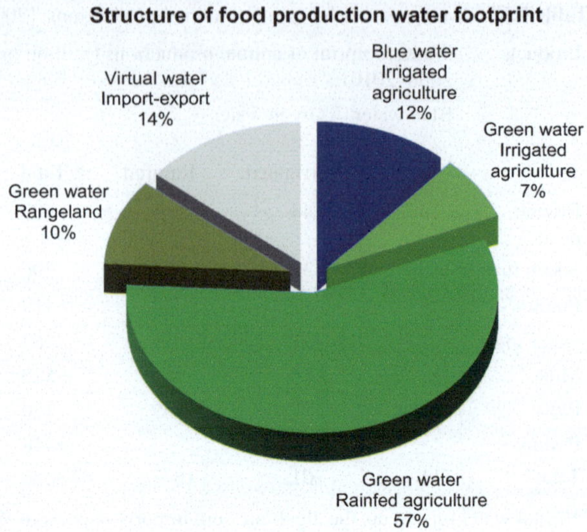

Fig. 7.11 Structure of the water footprint of food production (average 2006–2010)

7.5.2 Water Balance of Food Production

The WF of all food production represents an average potential of water resources amounting to about 12 km^3. These come from the water resources involved in the production of rainfed and irrigated crops, the WF of rangelands and import-export balance of fodder and cereals employed in animal feeding. The Table 7.14 shows the evolution of the different contributions (five-year averages) and highlights the structure of agricultural water balance of crop and animal productions.

Figure 7.11 shows the origin and nature of the water resources involved in food production during the period (2006–2010). It appears that agricultural production from rainfed agriculture represents an intake of 57% of water resources, the contribution of rangelands amounts to 11% and irrigated agriculture accounts for 19% (12% in BW and 7% in GW) of all the WF of total agricultural production. Overall, the GW contribution (including the rangelands) represents three quarters of the agricultural production. The remaining quarter corresponds for a half to the BWF associated to irrigation and for the other half to the contributions resulting from VW imports in the form of fodder and cereals for livestock production.

Regardless to the contribution of VW to the local animal production, the examination of the change in the various forms of WF of agricultural production indicates that the local water potential involved in agricultural production (BW and GW) has more than doubled over the past four decades following in this the increasing of the population which went from nearly 5.1 million in 1971 to over 10.5 million in 2010. This is a remarkable performance of the Tunisian agriculture which succeeded at keeping the specific agricultural production (WF per capita) stable despite the significant increase in population.

Table 7.14 Average water footprint of food production

Water footprint (million m^3)		1971–1975	1976–1980	1981–1985	1986–1990	1991–1995	1996–2000	2001–2005	2006–2010
Products of animal origin	Blue water	35	46	48	57	67	88	102	111
	Green water	1018	1332	1406	1661	1952	2543	2957	3226
	Virtual water	471	616	650	769	903	1176	1368	1492
	Total	1524	1994	2105	2487	2923	3807	4427	4829
Products of vegetal origin	Blue water	544	589	664	735	792	914	1021	1249
	Green water	3417	3309	3682	3426	4909	5299	5294	5686
	Total	3961	3898	4346	4161	5700	6213	6315	6935
Total food products	Blue water	579	635	712	792	859	1001	1123	1360
	Green water	4435	4641	5088	5088	6861	7842	8251	8912
	Virtual water	471	616	650	769	903	1176	1368	1492
	Total	5485	5892	6451	6648	8623	10,020	10,743	11,764

Data source Ministry of Agriculture databases, Ministry of Agriculture (2012)

It appears that the development of irrigated agriculture which contribution in BW rose from 0.58 to 1.36 km^3 has not fundamentally changed the structure of agricultural production, much of which continues to be provided by rainfed crops. For its part, rainfed agriculture has recorded significant improvements in its productivity. It also appears that the growth of food productions of animal origin has been faster than that of vegetable productions. Backed by imports of fodder, in significantly growing, the WF of the animal production sector has more than tripled between the early 1970s and the late 2000. Correspondingly, the part of water resources involved in the production of agrifood products of livestock origin has significantly increased from 28% in the early 1970s to 41% in the late 2000s. Consequently, the part of the foodstuffs of plant origin showed a decline of nearly 13%. This change in the structure of food production also reflects a change in the structure of the average diet in which the share of animal products is increased at the expense of foodstuffs of plant origin.

7.5.3 *Water Balance of Agrifood Trade*

The interpretation of trade exchanges, in terms of VW allows the calculation of the VW fluxes in order to assess their contribution to the overall water balance of the food demand. The analysis of the WF of the production of different crops has allowed to assess the specific WF of local animal and crop productions and to specify the origin and nature of water resources used in food production. The evaluation of VW fluxes associated with food products trade requires that we expand the list to foodstuffs which are not produced locally. We rely for this on data from literature and in particular the global data of Mekonnen and Hoekstra (2011). Table 7.15 presents the VW related to some imported food products.

Figure 7.12 shows the evolution of food production WF and the VW involved in foodstuffs imports and exports. In one way or another, the local production, highly variable due to the impact of climatic conditions on crop yields, especially that of rainfed crops, determines the VW fluxes: exports fluctuate together with the fluctuations of the productions and it is the same for imports that are a complement to local productions.

Figure 7.13 shows the average evolution of the VW related to imports and exports of food products (5-year moving average). Examination of this figure shows that the increase of the VW imports has been significant in recent decades going from less than 2 km^3 in the early 1970s to more than 8 km^3 at the end of the decade 2000–2010. During the same period, the exports of VW have doubled going from about 1 km^3 to over 2 km^3. These figures indicate that the water balance of the food trade balance is in deficit and the deficit has considerably amplified in recent decades. The coverage ratio of VW flows associated with agrifood trade that approached 80% in the early 1970s gradually deteriorated to stabilize at around 30% in the late 2000s.

The structure of the average VW involved in agrifood imports and exports during the period 2006–2010 shows the large fluxes, in both directions, of VW trade related

Table 7.15 Virtual water related to some imported food products

Products	Valuation basis references	Specific water footprint m^3/kg
Cheese	12 kg milk : 1 kg	15.96
Concentrated milk	2.2 kg milk :1 kg	2.936
Milk powder	8 kg milk : 1 kg	10.65
Colza oil	Mekonnen et Hoekstra (2011)	3.664
Rice	id	1.913
Spices	id	6.616
Soybean oil	id	4.117
Sunflower oil	id	6.387
Vegetable oil	id	5.818
animal fat	id	4.703
Alfalfa	id	0.134
Banana	id	0.757

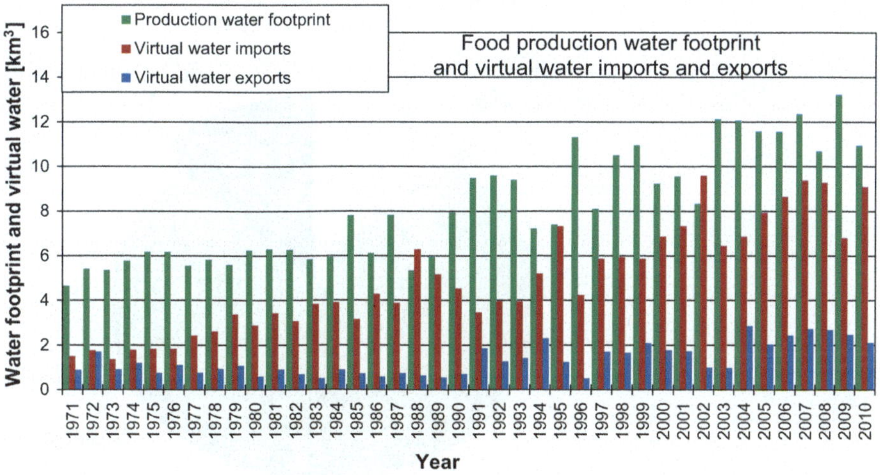

Fig. 7.12 Evolution of the food production water footprint and VW related to imports and exports of agrifood products

to cereals and edible oils exchanges. The olive oil represents 59% (1.47 km^3) of VW related to agrifood exports. Nevertheless, Tunisia imports large quantities of vegetable oil, which are superior to the VW related to olive oil exports, and which account for 18% (1.57 km^3) of the VW related to all imports foodstuffs.

The largest part of imports is related to cereals. It represents in VW, an intake of 3.19 km^3 or 37% of the VW fluxes associated with imports. This contribution is crucial in matching supply and demand for cereals whose local production is structurally insufficient. Cereal production for food consumption represents an average VW of approximately 3.48 km^3, slightly more than half the local demand.

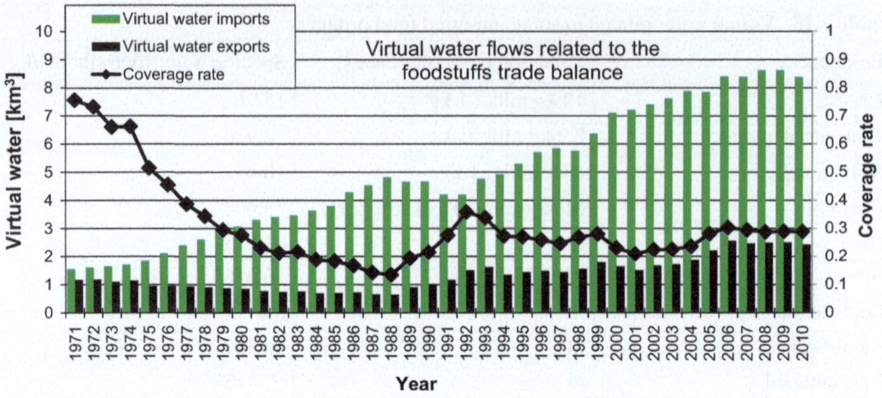

Fig. 7.13 Evolution of the average virtual water involved in imports and exports of agrifood products

Fig. 7.14 Structure of the virtual water exports

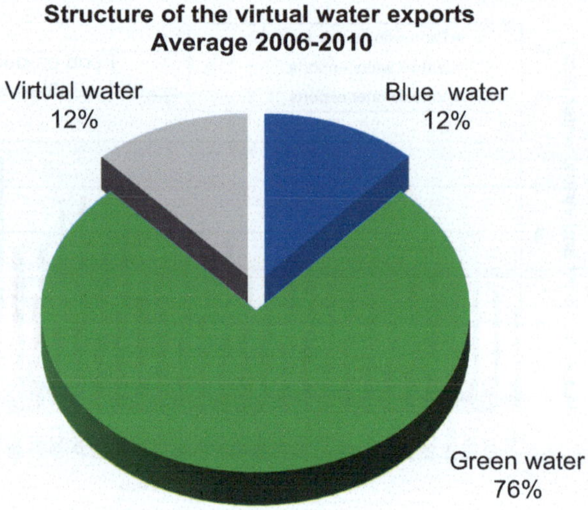

Figure 7.14 shows the nature of the VW related to exports of agrifood products. It indicates that the GW share is more than three quarters of the VW related to exports, emphasizing the major role of rainfed agriculture in the equilibrium of the food trade balance. However the contribution of irrigated area is not negligible. The share of BW represents 12% of VW exports, more than half of which comes from dates exports. It should be noted that 12% of exports come from VW flows associated with the importation of goods which are not produced locally (sugar, coffee, etc.), or from processing of imported products including basic products (cereals, fodder).

Table 7.16 Virtual water balance of the agrifood trade balance, (2006–2010)

Products	Virtual water related to agrifood exports and imports (million m^3/an—average 2006–2010)				
	Imports	Exports			
		Blue water	Green water	Virtual water	Total
Animal products	229	14	28		42
Oil and grease	1570	62	1409		1471
Leguminous products	115	0	5		5
Cereal grain-products	3191	5	228		233
Garden products	25	27	16		43
Fodder	1997	11	137		148
Fruits	58	173	69		242
Coffee, tea, spices	353			5	5
Sugar and sweets	1100			278	278
Wines and beverages	14	14	9		23
Total	8650	307	1901	283	2490

The main results related to the water balance of the foodstuffs trade balance, averaged over the period (2006–2010), are summarized in Table 7.16. The evolution of the various elements of these balance sheets over the past four decades is presented in Table 7.17.

The examination of the change in the water balance of the foodstuffs trade balance shows the significant increase in imports whose VW volume has increased more than five times between the early 1970s and the late 2000s. During the same period, the VW related to exports has increased by 2.5 times. It follows that the growth rate of VW imports was approximately twice that of exports and the water deficit has widened significantly from 560 million m^3 (about 100 m^3/capita/year) in the early 1970 to 6.16 Billion m^3 (approximately 600 m^3/capita/year) in the late 2000s.

Table 7.17 Evolution of the water balance sheet items related to the foodstuffs trade balance

Balance sheet items			1971–1975	1976–1980	1981–1985	1986–1990	1991–1995	1996–2000	2001–2005	2006–2010
Exports	Products of animal origin	Blue water	0	0	0	0	0	0	1	1
		Green water	1	2	3	1	5	9	16	28
		Virtual water	1	1	1	1	2	4	7	13
		Total	2	3	4	2	8	14	24	42
	Products of vegetal origin	Blue water	141	93	98	95	131	140	202	293
		Green water	925	728	627	525	1444	1358	1378	1873
		Virtual water	24	82	23	27	32	35	119	283
		Total	1090	903	748	647	1608	1533	1699	2448
	Total (million m^3)		1092	906	752	649	1616	1546	1723	2490
Imports	Products of animal origin		199	444	625	666	549	231	253	229
	Products of vegetal origin		1455	2186	2756	4167	4232	5535	7391	8421
	Total (million m^3)		1654	2630	3381	4834	4781	5765	7644	8650
Deficit	Products of animal origin		197	441	622	665	541	217	229	187
	Products of vegetal origin		365	1283	2008	3521	2625	4002	5692	5973
	Total (million m^3)		562	1724	2629	4185	3165	4219	5921	6159

References

Aldaya MM, Llamas MR (2009) Water footprint analysis (hydrologic and economic) of the Guadiana river basin within the NeWater project. Third edition of the United Nations World Water Development Report (WWDR-3)

Allan JA (1997) Virtual water: a long term solution for water short Middle Eastern economies? Paper presented at the 1997 British Association Festival of Science, Roger Stevens Lecture Theatre, University of Leeds, Water and Development Session—TUE.51, 14.45, 9 Sept 1997

Allan JA (1998) Moving water to satisfy uneven global needs. Trading water as an alternative to engineering it. ICID J 47(2):1–8

Allen RG, Pereira LS, Raes D, Smith M (1998) Crop evapotranspiration—guidelines for computing crop water requirements. FAO Irrigation and drainage paper 56

Besbes M, Chahed J, Bargaoui Z, Mattoussi MS, Mhiri A (2002) Water future, a new challenge for Tunisia. Int. Rep. Tunisian Institute for Strategic Studies, Tunis

Besbes M, Chahed J, Hamdane A, de Marsily G (2010) Changing water resources and food supply in arid zones: Tunisia. Water Sustain Arid Reg 103–121

Besbes M, Chahed J, Hamdane A (2014) Sécurité hydrique de la Tunisie, Gérer l'eau en conditions de pénurie. Ed. L'Harmattan, Paris

Chahed J, Hamdane A, Besbes M (2008) A comprehensive water balance of Tunisia: blue water, green water and virtual water. Water Int 33(4):415–424

Chahed J, Besbes M, Hamdane A (2011) Alleviating water scarcity by optimizing green virtual-water: the case of Tunisia. Value of water research report series no. 54, UNESCO-IHE, Delft, The Netherlands, 99–114

Chouchane H, Hoekstra AY, Krol MS, Mekonnen MM (2013) Water footprint of Tunisia from an economic perspective, value of water research report series no. 61, UNESCO-IHE, Delft, The Netherlands

Ercin AE, Mekonnen MM, Hoekstra AY (2012) The water footprint of France, value of water research report series no. 56, Delf (The Netherlands), UNESCO-IHE Institute for water education

Falkenmark M (1995) Land-water linkages: a synopsis. In: Mather T (ed) Land and water bulletin: land and water integration and River Basin Management, vol 1. Food and Agriculture Organization (FAO) of the United Nations, Rome, pp 15–16

Falkenmark M, Rockström J (2006) The new blue and green water paradigm: breaking new ground for water resources planning and management. J Water Resour Plan Manage Asce, 132(3)

FAO (2017) FAOSTAT: food and agriculture data. Statistics division. http://www.fao.org/faostat/en/#data

Henia L, Omrane MN, Hlaoui Z, Cherif A, Abderrahmen A (2008) Atlas de l'eau en Tunisie, Université de Tunis, Faculté des Sciences Humaines et Sociales

Hoekstra AY, Hung PQ (2002) Virtual water trade: a quantification of virtual water flows between nations in relation to international crop trade. Value of water research report series no. 11, Delf (The Netherlands), UNESCO-IHE Institute for water education

Hoekstra AY, Chapagain AK, Aldaya MM, Mekonnen MM (2011) The water footprint assessment manual: setting the global standard. Earthscan, London

Kampman DA, Hoekstra AY, Krol MS (2008) The water footprint of India. Value of Water Research, Report Series No. 32, UNESCO-IHE, Delft, The Netherlands

Karray B, Msallam M, Ksantini M, Mahjoub Boujnah D, Grati Kamoun N (2009) Programmes et acquis de recherch es pour la rénovation de la filière huile d'olive et l'amélioration de ses performances, Institut de l'Olivier – Tunisie

Liu J, Savenije HHG (2008) Food consumption patterns and their effect on water requirement in China. Hydrol Earth Syst Sci 12(3):887–898

MARH-DGGREE. (1995). Recueil des besoins en eau des cultures irriguées, Données de base. Rapport DGGR

Mekonnen MM, Hoekstra AY (2010) The green, blue and grey water footprint of crops and derived crop products. Value of water research report series no. 47, UNESCO-IHE, Delft, The Netherlands

Mekonnen MM, Hoekstra AY (2011) The green, blue and grey water footprint of crops and derived crop products. Hydrol Earth Syst Sci 15:1577–1600

Ministry of Agriculture (2012) Statistical agricultural yearbook for 2010. series from 2001 to 2010; DGEA, Min. agriculture, Tunis

Rezgui M, Ben Mechlia N, Bizid E, Kalboussi R, Hayouni R (2000) Étude de la stabilité du rendement de blé dur dans différentes regions de la Tunisie. In: L'amélioration du blé dur dans la région méditerranéenne: nouveaux défis. Options méditerranéennes, Série A: Séminaires Méditerranéens, vol 40, pp 167–72

Rezgui M, Zairi A, Bizid E, Ben Mechlia N (2005) Consommation et efficacité d'utilisation de l'eau chez le blé dur (Triticum durum Desf.) cultivé en conditions pluviales et irriguées en Tunisie. Cah Agricultures 14(4):391–397

Salmoral G, Aldaya MM, Chico D, Garrido A, Llamas MR (2010) The water footprint of olive oil in Spain, Papeles de agua virtual, Observatorio del Agua, Numéro 7, Editon: Fundación Marcelino Botín. Pedrueca, (Santander), www.fundacionmbotin.org

Schyns JF, Hoekstra AY (2014) The added value of water footprint assessment for national water policy: a case study for Morocco. PLoS ONE 9(6):e99705

Van Oel PR, Mekonnen MM, Hoekstra AY (2009) The external water footprint of the Netherlands: Geographically-explicit quantification and impact assessment. Ecol Econ 69(1):82–92

Vincent D, De Caritat AK, Stijn Bruers S, Chapagain A, Weiler P, Laurent A (2011) Belgium and its water footprint. WWF-Belgium, Brussels

Chapter 8
Water Security, Food Security and the National Water Dependency

Contents

8.1 Introduction

At the current state of our knowledge, reaching and sometimes exceeding the exploitability of conventional water resources has become a reality. It becomes increasingly difficult to continue the development of large-scale irrigated agriculture without consequences on water and soil resources sustainability. In these conditions, increasing allocations of direct water uses, domestic and industrial, can only be made to the detriment of agricultural use. Agricultural allocation is therefore condemned in the long term to decrease at the rate of the demand increase from the other sectors: how, in that context, does the future of water in Tunisia look?

The long-term water security covers the security of water supply, but also food security since agriculture accounts for a large share of water resources: how to meet the growing needs of the different sectors of the economy without weakening the farming sector and the country's food security?

M. Besbes et al., *National Water Security*,
https://doi.org/10.1007/978-3-319-75499-4_8

Without providing definitive answers to this last question which has largely determined Tunisia's water policy, outlook of water-balance future evolution is presented in a water-food overview including all kinds of water resources: withdrawals (blue water, BW), water footprint of rainfed cultivated areas and rangelands products (green water, GW), water contribution of import-export food balance (virtual water, VW). This prospective exercise identifies the key factors that control water future, and prepares guidelines to assist in the water dependency control as part of comprehensive approaches for the development of food security policies. Indeed, food security goes beyond the simple equilibrium of agrifood products trade balance to cover different themes which should be examined in an integrated perspective combining agronomic, economic, environmental, social dimensions (Bachta and Zaïbet 2006), but also educational, scientific and technological perspectives (Besbes et al. 2002).

The food demand balance is interpreted in the form of a Water Dependency Index that represents the ratio of net VW imports to the total food demand water requirements. Comprehensive water resource assessment and long term projection of the Water Dependency Index (WDI) allow the assessment of the water resources potential, and identify ways to develop all forms of water resources in order to increase national potential of water demand satisfaction and strengthen food safety. This poses new challenges for irrigated agriculture in Tunisia and generally in water scarce countries, which should improve their productivity by improving the water uses efficiencies and leads to the obvious conclusion that water scarce countries have to rely on all their resources to develop and support their agricultural development efforts.

8.2 National Water Balance: The "Water Dependency Index"

8.2.1 Virtual Water, Food Trade and Food Security

Food trade and food security

In Tunisia, as in all arid regions, agricultural intensification has emerged as a great way to increase production and to meet the food challenge. Vast water resource mobilization programs have been carried out and various incentives and support measures have been put in place to promote irrigation. The central fact is that irrigation consumes a lot of water and its development depends first and foremost on the availability of water and soil resources. The stabilization of the water supply and the prospect of reallocation of water resources to the detriment of agricultural sector represent a key question of the water issue in Tunisia. The aim is to find the most appropriate ways to meet the future growing needs of other water use sectors without straining the agricultural sector.

The production potential of all water and soil resources is insufficient to meet the food demand, ever increasing in quantity and quality. Therefore, imports and exports of foodstuffs products are needed in order to balance supply and demand. The crucial challenge for Tunisia is to optimize these VW exchanges, valuing the comparative advantages that local agriculture can develop and anticipating impacts of any changes that may occur nationally and internationally, which could cause terms-of-trade deteriorations: market liberalization, changes in agricultural and food subsidies policies, etc.

Irrigated agriculture has occupied a prominent place in the strategy of trade enhancement in agrifood products. The two flagship products of irrigated products exports are citrus fruits and dates. Exports of dates have changed significantly in quantity and value over the past fifteen years; they account for more than the half of export earnings from fruits and vegetables, to a large extent produced by irrigated crops. Dates export earnings have increased from less than US$40 million in 2000 to more than US$230 million in 2013. Export earnings from citrus fruit, relatively smaller, have not evolved at the same rate, going from about US$7 million in 2000 to US$11 million in 2013. However, the development for irrigation strategy has not significantly changed the trade balance structure of food products, which is still dominated by rainfed crops productions. Olive oil, almost entirely produced within rainfed groves, represents the main agricultural product exported by the country. Earnings from olive oil exports have increased from less than US$200 million in the early 2000s to more than US$300 in the early 2010 and reached more than US$500 million in 2013. More generally, rainfed agriculture plays an essential role in food security; its role in the food balance equilibrium is twofold. On the one hand, production of rainfed agriculture constitutes an important part of food products exports. On the other hand, imports of basic commodities including cereals, which are an important item of food imports, are intended to cover the shortfall in local production mainly cultivated under rainfed conditions and whose performance is highly dependent on climatic constraints. This issue is all the more important as the wheat imports bill has increased considerably during the last decade. National expenditure on wheat imports has been raised from less than US$200 million/year in the early 2000s to more than US$500 million/year in the early 2010 and reached more than US$800 million/year in 2008.

Developing grain reserves as virtual water storage

In Tunisia, cereals needs are around 5 million tons per year. The consumption of wheat in 2012 is estimated at 2.25 million tons corresponding to about 4.5 km^3 water requirements, i.e. one and half times the total abstracted country's BW. Thus, constituting strategic grain reserves for annual wheat requirements means to create water reserves equivalent to more than two times the surface water storage capacity of the overall hydraulic infrastructure estimated at 2.1 km^3.

Storage of agricultural production surpluses in wet years is a form of water resources storage which can come from blue and GW resources. Similarly, storage of imported foodstuffs represents a kind of VW storage. This correspondence between food reserves and water reserves is relevant: it allows the interpretation of the foodstuffs stockpile capacity in terms of water resources storage capacity. It is also quite realistic since water reserves are in fact also used to irrigate cereals, especially as complementary irrigation. The same reasoning can be extended to other agrifood products, especially to olive oil produced mainly by rainfed crops.

The current national grain storage capacity is estimated at 6.3 million quintals (Office of Cereals 2017). With the planned storage infrastructures, the national storage capacity should be extended to 6.4 months of wheat consumption (World Bank 2012). This would be equivalent to a VW storage of approximately 2.4 km^3. Similarly, olive oil storage capacity amounting to 365,000 tons (National Oil Office, 2017), would be equivalent to a VW storage of over 3 km^3. Taken together, the projected storage capacities of cereals and olive oil represent the equivalent of more than two and a half times the storage capacity of surface BW of the country. As the national production of cereals and olive oil come mainly from rainfed agriculture, the grain silos and olive oil containers storage capacities would be equivalent to a GW storage capacity of a 5.4 km^3.

In this perspective, the foodstuffs storage is a convenient form of water resources storage which is inexpensive and without environmental consequences. In Tunisia, the storage cost of one ton of grain amounts to US$42/year. Converted to water storage cost, this corresponds to a cost of 2.1 cents per m^3 of stored water, paltry in comparison with storage by dams. Moreover, wide improvement margins are possible because the annual cost of storing grain in Tunisia is among the highest: it is only US$25 in USA, US$22 in Morocco and around US$15 in South Korea (World Bank 2012).

These figures show that Tunisia has interest in developing and refining its production strategies, procurement and management of food stockpiles. In addition to their strategic aspect, food reserves can help regulate stocks to manage droughts. Drought management requires establishment of a comprehensive and coherent program to develop and promote rainfed farming and pastoral activities across the country. Ongoing programs are in line with this objective: the Tunisian State intends to increase the national grain storage capacity. These efforts should continue as part of a comprehensive and coherent program to develop and promote rainfed crops across the country.

For olive oil, the storage capacity is already large enough to absorb the disturbances associated to annual production variability. Estimated at 365,000 tons, this capacity represents two times the national average olive oil production, which is estimated at 180,000 tons per year. However the olive production is subject to important inter-annual variability, so olive oil production reached the exceptional level of 310,000 tons in 1996/1997 and fell to 35,000 tons in 2001/2002. Considering the biannual alternate-bearing cycle of olive trees productions, the storage capacity equivalent to two years average production appears convenient for olive oil sector management and for the orderly marketing of the production.

8.2.2 *The Comprehensive Water Balance of Food Needs*

Balance sheets of the agrifood trade

Increase in food demand is generally accompanied by a significant increase of VW imports. Figure 8.1 shows the variations (moving average over 5 years) of VW coverage rate (proportion of exports to imports) related to external food trade. In terms of virtual water balance, the coverage rate of the foodstuffs imports by exports had deteriorated: it dropped from near 80% in the early 1970s to about 30% in the 2010s.

But at the same time, the food trade balance is very favorable to Tunisia: in 2010, the coverage rate was 30% in VW volumes, but near 80% in value (Fig. 8.1). The revenue from exports of food products (expressed in VW) is more important than spending on imports of foodstuffs of equivalent VW volume.

Table 8.1 specifies the annual average flux of VW over the period 2006–2010 and the financial flows associated with for traded food products. The benefits derived from commercial exchange of food products are due to terms-of-trade which, as regards to water, are favorable for Tunisia in particular concerning the exports of olive oil and agrifood products that have industrial added value (cereal products, fruits, canned products…).

Water requirements for national food demand

By evaluating the water footprint of agricultural production and VW fluxes related to food products trade, it becomes possible to determine the water balance of food needs. Table 8.2 shows the evolution of the water requirements for food needs broken down according to their origin and nature and presented on successive five-years averages. Calculation results show that water requirements for food demand has

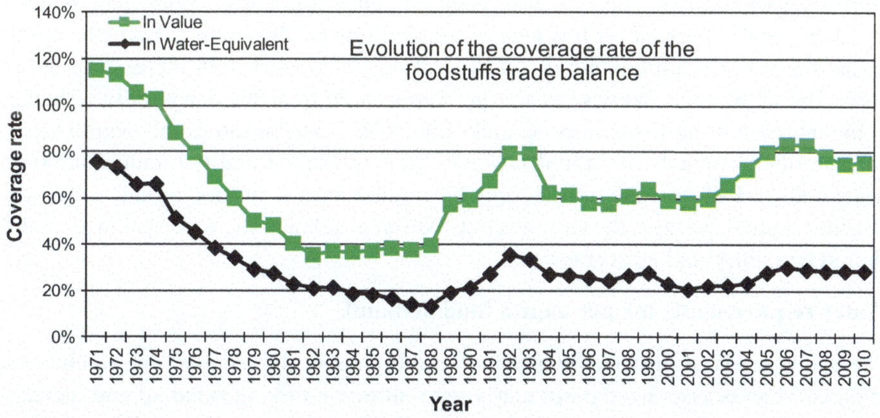

Fig. 8.1 Coverage rate of the foodstuffs trade balance (ratio of exports to imports), from 1971 to 2010

Table 8.1 Trade balance of main foodstuffs in terms of VW and value, (five-year average: 2006–2010; 1 DT eq. US$0.694 in 2010)

Products	Imports			Exports		
	Virtual water Mm^3	Value MDT	Value 10^{-3} DT/m^3	Virtual water Mm^3	Value MDT	Value 10^{-3} DT/m^3
Products of animal origin	229	89	389	42	42	980
Oil and grease	1570	377	240	1471	837	569
Leguminous products	115	12	101	5	<1	140
Cereals and cereal products	3191	999	313	234	131	559
Vegetables, canned products	25	47	1929	43	81	1894
Fodder	1997	137	68	148	27	184
Fruits and fruit juices	58	37	636	242	275	1139
Coffee, tea, spices	353	64	181	5	24	4622
Sugar and sweets	1100	246	223	278	56	201
Wines and beverages	14	13	933	23	54	2345
Total	8650	1972	228	2490	1528	614

Source "Author's elaboration and Statistical yearbooks of Ministry of Agriculture, Tunisia"

more than tripled over the past four decades. They amount to more than 16 km^3 in 2010. As the population has nearly doubled during the same period, the result is that the water requirements for individual dietary recorded an increase of about 50%. The significant increase of the per capita food demand is naturally reflected in the increase of agrifood imports and of the VW contribution to the overall water balance of food needs. It should however be emphasized that the annual national water budget does not take into account the regulatory role of food products storage. On inter-annual average, the storage contribution is null and the water balance model should be interpreted as average.

Water requirements for per capita food demand

The water needs of agriculture and food are high and may change due to changes in dietary habits associated particularly with improving living standard and increasing purchasing power of the population. Individual dietary needs have registered a significant increase, both for vegetal and animal products, with a larger increase for poultry meat and dairy products. The water requirements for per capita food

Table 8.2 Average annual water requirements for food needs (km^3/year)

Five-year average		1971–1975	1976–1980	1981–1985	1986–1990	1991–1995	1996–2000	2001–2005	2006–2010
Products of animal origin	Blue water	0	0	0.1	0.1	0.1	0.1	0.1	0.1
	Green water	1.0	1.3	1.4	1.7	1.9	2.5	2.9	3.2
	Virtual water	0.7	1.1	1.3	1.4	1.4	1.4	1.6	1.7
	Total	1.7	2.4	2.7	3.2	3.5	4.0	4.7	5.0
Products of vegetal origin	Blue water	0.4	0.5	0.6	0.6	0.7	0.8	0.8	1.0
	Green water	2.5	2.6	3.1	2.9	3.5	3.9	3.9	3.8
	Virtual water	1.0	1.5	2.1	3.4	3.3	4.3	5.9	6.6
	Total	3.9	4.6	5.7	6.9	7.4	9.0	10.6	11.4
Total food needs	Blue water	0.4	0.5	0.6	0.7	0.7	0.9	0.9	1.1
	Green water	3.5	3.9	4.5	4.6	5.4	6.5	6.9	7.0
	Virtual water	1.6	2.5	3.4	4.8	4.7	5.7	7.5	8.4
	Total	5.6	7.0	8.4	10.1	10.9	13.1	15.3	16.4

demand had grown from about 1000 m^3/year/capita in the early 1970s to nearly 1600 m^3/year/capita in 2010. It is worth recalling that water resources allocated to irrigation sector (BW) which represent, in the late 2000s, 80% of national water resources, represent only a per capita allocation of about 200 m^3/year/capita, or about one eighth of the food demand water requirements. The implications of such changes on water balances and agriculture are considerable, especially as the future increased allocations for direct water uses (municipal and industrial) can only be made at the expense of agricultural use.

8.2.3 The Water Dependency Index (WDI)

The core idea behind the comprehensive water balance model proposed and adjusted in previous works (Chahed et al. 2007; Besbes et al. 2010), is to consider that the water allocated to irrigation (IW) is the quantity of BW available when the direct water demand (DD: Industry, Tourism, Communities) has been satisfied (Eq. 8.1). In this model, virtual water (VW) expresses the water requirements balance of food demand (FDWR), considering the amounts of water involved in rainfed crop production (GW) and irrigated agricultural production (Eq. 8.2).

$$\mathrm{IW} = \mathrm{EWR} - (1 - \mathrm{RI})\mathrm{DD} \tag{8.1}$$

$$\mathrm{VW} = \mathrm{FDWR} - \mathrm{GW} - \mathrm{k}\,\mathrm{EWR} + \mathrm{k}(1 - \mathrm{RI})\mathrm{DD} \tag{8.2}$$

IW	Irrigation Water volume
EWR	Exploitable Water Resource
RI	Recycling Index
DD	Direct water Demand
VW	Virtual Water volume
FDWR	Food Demand Water Requirements
GW	Green Water volume
k	Irrigation factor (converts irrigation volumes into water footprint of the irrigated food production; this factor integrates irrigation efficiency and rainfall contribution).

The model verification consists in testing values of the parameters k and RI, while providing data concerning EWR, DD, FDWR, then calculating the virtual water volume VW. The calculated VW should correspond to the observed import-export balance. We however must note that the model is strongly constrained because the total food demand water requirements has already been estimated taking into account the net contribution of the import-export balance; it is therefore not possible to do any further calibration of the model but simply to verify the results consistency. After checking, the model is very useful to the extent that it may be applied prospectively.

The food demand water requirements (FDWR) include the water-equivalent of agricultural production (PRD) consumed on the local market (PRD-EXP) and the water-equivalent of agrifood imports (IMP):

$$\mathrm{FDWR} = \mathrm{PRD} - \mathrm{EXP} + \mathrm{IMP} \tag{8.3}$$

The water-equivalent of agricultural production (PRD) includes green water (GW) and blue water (BW). The latter can be evaluated using a factor (k) that integrates irrigation efficiency (BW = kIW), Eq. (8.4). The volume allocated to Irrigation (IW) is calculated as the remaining exploitable water resources net of direct demand (DD) and environmental demand (ENV). The exploitable water resources (EWR) include surface water (SFW), groundwater (GDW) and volumes produced by desalination (DSL). The share of resources allocated to direct uses (industry, tourism, collectivities) is recycled at the overall rate (RI).

$$\mathrm{PRD} = \mathrm{GW} + \mathrm{k}[\mathrm{SFW} + \mathrm{GRW} + \mathrm{DSL} - (1 - \mathrm{RI})\,\mathrm{DD} - \mathrm{ENV}] \tag{8.4}$$

The comprehensive food-demand water balance model indicates that the water security of a nation should ideally refer to the total sum of water resources and water uses to face, at the best of national possibilities, to the food challenge. The model takes into account all components of water resources: (i) available water resources mobilized from surface and ground water systems; (ii) soil water resources involved in food production (crops production, pasture); (iii) water resources produced by unconventional methods (treated waste water reuse, desalination); and (iv) VW flows related to international trade. The first two components can be considered as part of the water cycle. Along with the third component, this represents what is produced and used locally. The fourth component of comprehensive water resources represents the cross border flux of VW which balance represents the difference between all the water requirements for all population needs and all the water footprint of all national production. The "Water Dependency Index" (WDI) defined by Eq. (8.5) represents thus the net VW volumes within the total food demand.

$$\mathrm{WDI} = \frac{\mathrm{IMP} - \mathrm{EXP}}{\mathrm{FDWR}} = 1 - \frac{\mathrm{GW} + \mathrm{k}\,[\mathrm{SFW} + \mathrm{GDW} + \mathrm{DSL} - (1 - \mathrm{RI})\,\mathrm{DD} - \mathrm{ENV}]}{\mathrm{FDWR}} \tag{8.5}$$

Table 8.3 shows the evolution of VW balance and specifies some aspects of the food dependency and water dependency of Tunisia. Expressed in water-equivalent, the contribution of local production to food needs has generally deteriorated and the overall water dependency index (WDI) has increased from 10% in the early 1970s to nearly 40% during the decade 2000–2010, (Table 8.4). On the other hand, Table 8.3 shows that the VW balance deficit related to products of animal origin has been reduced, slightly in volume and significantly when calculated on a per capita basis. It has been mentioned in Chap. 7 that the decrease in imports of agrifood products of animal origin which began in the late 1980s was accompanied by a significant amplification of animal feed imports to support the local livestock production. It is as if Tunisia has replaced the importation of livestock products and foodstuffs of animal origin by the importation of livestock feeds. In addition to the obvious benefits that could come up from such conversion in terms of food security (livestock products traceability, sanitary and veterinary control, etc.), it is also understood that by this mean, Tunisia has sought to stimulate its livestock sector and to value the production and commercialization chains. It would be useful, in this context, to assess the socioeconomic impacts of these efforts and to know, for example, the long term effectiveness of such policies especially as Tunisia began, since early 2000s, to face milk over-production to such an extent that there were concerns in the sector about the management of surplus production, (Khaldi and Naili 2001).

8.3 Food Security Versus Water Dependency: A Prospective Exercise

8.3.1 Water Footprint of Food Demand

Findings and order of magnitude

The overall results (Table 8.5) show that the largest share of mobilized water is intended to irrigated agriculture, the largest share of agricultural production is provided by rainfed agriculture and a significant part of the food demand is provided by food trade balance. On the period 2006–2010, the average water footprint (WF) of all food production in Tunisia amounts to approximately 12 km^3/year. This resource comes from water resources involved in crop production (rainfed and irrigated agriculture), WF of natural pastures and VW coming from the net import-export balance sheet of fodder and cereals for animal nutrition. The share of crops amounts to nearly 60% of the WF of food production and the livestock production represents approximately 40%.

Furthermore, our estimates indicate that rainfed agriculture accounts for 56% of WF of agrifood production, the rangelands contribution is 11% and irrigated agriculture provides 19% (12% BW and 7% GW) of the whole WF of food production. Overall, the contribution of GW (including rangelands) represents three-quarters of food production. The remaining quarter represents for about half the VW intake resulting from imports of fodder and grain for livestock production; and for the other half to the BW associated with irrigation. We should remember that all BW allocated to irrigation, 80% of mobilized water resources, contributes to 12% of the WF of national agricultural production

It is also interesting to note that estimations of BW footprint (BWF) of irrigated production results in a value of 1.4 km^3/year. If these "net water consumption" are applied using the whole hydraulic systems efficiency estimated at 60%, we

Table 8.3 Population and water requirements for food demand, five years average

Water requirements for food demand									
Five-year average		1971–1975	1976–1980	1981–1985	1986–1990	1991–1995	1996–2000	2001–2005	2006–2010
Population (×1000 inhab.)		5335	6046	6894	7738	8572	9333	9840	10,329
Food of animal origin	Water req.[a]	1.7	2.4	2.7	3.2	3.5	4.0	4.7	5.0
	Spec. water req.[b]	323	403	395	407	404	431	473	486
Food of crop origin	Water req.	3.9	4.6	5.7	6.9	7.4	9.0	10.6	11.4
	Spec. water req.	723	755	827	893	866	968	1081	1105
Total food demand	Water req.	5.6	7.0	8.4	10.1	10.9	13.1	15.3	16.4
	Spec. water req.	1045	1158	1223	1301	1270	1400	1554	1591

[a]Water requirements (km^3/year)
[b]Specific water requirements on a per capita basis (m^3/year/capita)

Table 8.4 Evolution of water requirements for food needs and water dependency index

Virtual water balance and water dependency index (WDI)

Five-year average			1971–1975	1976–1980	1981–1985	1986–1990	1991–1995	1996–2000	2001–2005	2006–2010
Population [×1000 inhab.]			5335	6046	6894	7738	8572	9333	9840	10,329
Imports	Food of animal origin	Virtual water[a]	199	444	625	666	549	231	253	229
		Per capita VW[b]	37	73	91	86	64	25	26	22
	Food of vegetal origin	Virtual water	1455	2186	2756	4167	4232	5535	7391	8421
		Per capita VW	273	362	400	539	494	593	751	815
Exports	Food of animal origin	Virtual water	2	3	4	2	8	14	24	42
		Per capita VW	0.4	0.5	0.6	0.3	0.9	1.5	2.4	4.1
	Food of vegetal origin	Virtual water	1090	903	748	647	1608	1533	1699	2448
		Per capita VW	204	149	109	84	188	164	173	237

(continued)

Table 8.4 (continued)

Virtual water balance and water dependency index (WDI)										
Five-year average			1971–1975	1976–1980	1981–1985	1986–1990	1991–1995	1996–2000	2001–2005	2006–2010
Deficit	Food of animal origin	Virtual water	197	441	622	665	541	217	229	187
		Per capita VW	37	73	90	86	63	23	23	18
	Food of vegetal origin	Virtual water	365	1283	2008	3521	2625	4002	5692	5973
		Per capita VW	68	212	291	455	306	429	578	578
Water dependency index (WDI)			0.10	0.25	0.31	0.42	0.29	0.32	0.39	0.37

[a]Virtual water (million m^3)
[b]Per capita virtual water (m^3/year/capita)

Table 8.5 Summary of agricultural productions water footprint (2006–2010)

Agricultural production	Agricultural production water footprint (km^3/year) Average 2006–2010						
	Blue water	Green water				Virtual water	Total
		Irrigated	Rainfed	Pasturage	Total		
Crop productions	1.3	0.7	5		5.7		7
Animal productions	0.1	0.2	1.7	1.3	3.2	1.5	4.8
Total	1.4	0.9	6.7	1.3	8.9	1.5	11.8

Table 8.6 Water balance sheet of the food demand

Origin	Water balance sheet of the food demand (km^3/year)				
	Average 2006–2010				Average 2011–2013[a]
	Blue water	Green water	Virtual water	Total	
Production	1.4	8.9	1.5	11.8	10.5
Imports			7.1	7.1	10.5
Exports	0.3	1.9	0.3	2.5	2.9
Balance	1.1	7	8.3	16.4	18.2
Population (×1000 inhabitant)				10,329	10,779
Specific water demand (m^3/year/capita)				1591	1690
Water dependency index				0.37	0.42

[a]Data from FAOSTAT (Consulted on 15 August, 2017)

retrieve with a remarkable precision, the volume used by irrigation during the period 2006–2010 which amounts to 2.2 km^3/year.

To supplement its food needs, Tunisia imports in the late 2000s a total VW of 8.6 km^3/year in the form of food products (cereals, fodder, vegetable oil, sugar etc.). At the same time, Tunisia exports agricultural products (citrus, dates, vegetables, olive oil etc.) equivalent to 2.5 km^3 per year, leaving a balance of 6.1 km^3 (over a third of the all food needs). The net contribution of VW is partly (4.6 km^3) associated with balance of trade balance of foodstuffs; and the rest (1.5 km^3) corresponds to the contributions in VW associated with fodder imports to support local livestock production. These figures (Tables 8.5 and 8.6), with orders of magnitude they imply, indicate that the issue of water security may be set at nationwide in connection with the critical issue of food security and its relationship with agricultural production and of the food trade balance.

Future challenges in terms of water resources

Number of studies have already established projections on the future of blue water resources in Tunisia. These studies have proposed resource-use confrontations that can close the balance sheets at different horizons (GEORE 2001; Louati et al. 1998; ITES 2011a). Today, the mobilization of surface water and groundwater has already reached a very high level and the situation is such that simply maintaining the current mobilization potential requires that we solve the groundwater overexploitation problem and compensate dams' storage capacity losses from silting. Indeed, many dam reservoirs are subject to varying degrees to siltation and the phenomenon is amplifying and generalizing. Siltation, which reached nearly 500 million m^3 in 2010, is expected to reach nearly 1.5 km^3 in 2050, or more than 70% of the total surface water storage capacity (Ben Mammou and Louati 2007). The protection of water infrastructure, rehabilitation or simply their renewal or replacement cover important issues, not only to safeguard the potential of current mobilization but also to ensure that the water projects continue to play roles, often multiple and sometimes incompatible for which they were built (storage, irrigation and groundwater recharge, for example),. To achieve these objectives, Tunisia will be called in the coming decades, to design and implement large-scale hydro development programs.

With increased mobilization and increased pressure on the resource, the use of non-conventional water will be an alternative more and more realistic to the problem of water scarcity. In the area of reuse of treated wastewater, the effective utilization of treated wastewater is still low. As for desalination, which has grown significantly, especially during the last decade, recent studies suggest that it will be in the future an inevitable option for the supply of drinking water especially in central and southern regions, due to the pressure becoming stronger on conventional resources and constraints on the water quality. The actual production capacity of desalinated water is around 40 million m^3/year in 2010 and projections of SONEDE foresee that the production of water by desalination will reach 120 million m^3/year in 2030.

Chapter 7 has developed an approach to estimate the WF of rainfed and irrigated agricultural production. It appears that during the period 2006–2010, the average green water footprint (GWF) of agricultural products amounted to about 9 km^3/year the largest part of which comes from rainfed crops. Rainfed agriculture, with its various components, occupies an area of 4.5 million hectares and this area has changed very little over the past four decades. The report Eau 2050, in its Annex 3 (ITES 2011b), presents the future of rainfed agriculture, specifying the various difficulties and opportunities of the sector. In general, it seems that the potential of rainfed agriculture is great and the report proposes strategic directions with a concrete action program for the development of this potential. The study also calls for action programs to improve the recovery of rain water by rehabilitating traditional techniques, strengthening the consolidation of agricultural land in order to achieve optimal of farm sizes and promoting price policies of the main products of rainfed agriculture.

8.3.2 *Outlook for Blue Water Resources*

Conventional water resources

The current context can be described by the essential aspects that characterize today our knowledge of the resource, its potentialities, and the ways of using and managing water, but also by a number of important tendencies which prolong what seems to characterize permanently the evolution of the predominant socio-economic and human factors which will condition in the future, directly or indirectly, the management and use of the resource. At the current state of our knowledge on water resources, the future of BW is shaped by three hallmark features: (i) continuing to mobilize conventional water by collecting flows that still escape mobilization; (ii) strengthening mechanisms that support demand-side management and efficiency of water uses; (iii) supporting sustainable non-conventional water resources development and uses. Much has already been done in the area of conventional water resources development and management. Current surface water management programs are going further in consolidating surface water resources mobilization and interconnecting hydraulic infrastructure. As for groundwater, the exploitation of aquifers has reached such high levels that it becomes necessary to scale down the water abstractions in some regions to solve aquifers overexploitation and salinization problems. The future programs are also part of the same approach: mobilizing all available water resources and ensuring effectiveness of their uses.

Non-conventional water resources

While the development and management of conventional water resources are guided according to long-term vision, it would appear that the strategy of non-conventional water resources is less clear either as regards treated wastewater reuse or as regards sea water desalination for which there are fast evolutions, not necessarily compliant with previous projections and programs.

Number of studies carried out over the last twenty or thirty years had expected that non-conventional resource development potential would mainly concern treated wastewater reuse for which long term projections were targeting an almost full wastewater recycling. According to these studies, the other methods of production of non-conventional water resources were alleged to face great difficulties to compete economically with conventional resources. In particular desalination had been expected to remain limited to drinking water supply in remote areas that are lacking good quality water resources. As a result, the long-term projections had very little ambition for the development of seawater desalination (Louati et al. 1998). The objectives of the national programs on non-conventional water resources were largely based on these studies findings.

Despite the intense efforts made over the last three decades in promoting treated wastewater reuse, many problems are currently still hampering its massive expansion, especially for agricultural purposes. In return, we see the past few years an enormous growth in desalination projects. While the initial desalinated water production forecast for 2030 was 49 million m^3 (Louati et al. 1998), more recent projections by SONEDE (Zaara 2008) envisaged that water production by desalination will reach 120 million m^3 by 2030. Furthermore, the severe repetitive disruptions of water supplies observed over the last few years have created a perception of water shortage. This perceived shortage of water that infused media and official discourses has reinforced an idea already present that seawater desalination will become necessary to secure drinking water supply. This is in stark contrast to principles, or fundamental factors on which drinking water supply services have been based. Consequently, desalination programs have been considerably amplified and accelerated. In this regard, the final adoption of seawater desalination plant to be built in Sfax (100,000 m^3/day expandable to 200,000 m^3/day) constitutes a decisive turning point. Several other projects are currently being developed or implemented for water quality improvement in Southern Tunisia with a desalinated seawater production capacity of 67,200 m^3/day, like Djerba and Zarrat seawater desalination plants (50,000 m^3/day each respectively expandable to 75,000 and 100,000 m^3/day).

The newly implemented projects and their scope will overturn the structure of drinking water system based so far on conventional water resources transferred from region with water surpluses to deficit regions. According to the principles which govern water resources allocations, drinking water supply is a top priority and the traditional approach to resource-use confrontation is to prioritize the allocation of good quality water to drinking water supply. Consequently, the allocations increase to meet drinking water demand will be performed at the expense of agricultural use. Given that the agricultural sector accounts for the largest share of water resources, resource reallocation should, in principle, protect the supply of drinking water from shortages.

The SONEDE ongoing projects will totalize a nominal production of desalinated seawater of 123 million m^3/year expandable to more than 163 million m^3/year. The completion of the current projects will immediately bring production potential of desalinated water up to 168 million m^3 (expandable to 208 million m^3) as against the 45 million m^3 in 2011. Desalination is no longer just a mean of achieving specific goals (supply of isolated remote settlements or activities, improvements in the quality of public water supplies). It is becoming a main water supply source indispensable to provide entire regions and important metropolises with drinking water. It is also to be expected that industrial and touristic water demand will undergo similar transformations leading to substantial increases in seawater desalination. Alone, the SONEDE desalinated seawater production potential will, in the near future, represent almost the quarter of the total drinking water allocation and this part is set to rise maybe even exceeding the third, in the longer term (SONEDE 2017).

This is a turning point in fundamental principles underlying drinking water supply, which will clearly have technical and financial implications. The financial implications of these profound transformations have not been addressed in detail but one may expect that they will be crucial. National organisms having in charge of water services will have to review their pricing systems, and practices to preserve the basic financial balance taking into account all the technical, economic, social and environmental constrains. The supply of water is based on a principle of national solidarity: inter-regional transfer of conventional water resources, and adequate sharing of financial burden faced by consumers. The pricing system for drinking water adopted by the Tunisian authority, which provides for basic water services at affordable prices throughout the country with uniform nationwide tariffs, will certainly alleviate the charges faced by consumers due to the huge increase in desalination, but what remains clear is that the costs of providing/distributing drinking water will inevitably expand. That the supplementary expenses due to desalination could not be directly passed on consumers, can lead to inefficiencies in use of precious resource.

Population projections and drinking water demand evolution

The assessment of future drinking water demand is based on forecasts of future population and its water demand. The United Nations have developed probabilistic projections of the population evolution for all countries. The projections are presented for medium, high, low and constant-fertility variants and reviewed periodically. Recent findings suggest that the significant deceleration of fertility rates that marked the last decades of the last century in the North Western African countries has slowed (Ouadah-Bedidi et al. 2012). These authors have shown that fertility remains above replacement level in all North West African countries. In Tunisia, the total fertility rate (TFR), after bottoming out at 2.0 children per woman in the mid-2000s, has since witnessed slight growth during the 2010s, stabilizing at 2.3-2.4 children per woman during the 2010s, (INS 2017). The fertility decline appears to have leveled off at above replacement level.

The stabilization of total fertility rate has already started making impact on population projections. For instance, the Population Reference Bureau (2016) adopted demographic forecasts for 2050, close to the 2015 UN population projections with high variant hypothesis. Moreover the regular updating of the population projections appears to include the recent trend of fertility stabilization and the future population size has been continuously revised upwards. The outcomes of long term projections to 2050 and 2100 related to 2010, 2015 and 2017 revisions (UN 2011, 2015, 2017) are presented in Table 8.7.

Projections, with different assumptions about the population evolution, provide varying results from one assumption to the other. According to the medium variant hypothesis, the last forecast (The 2017 revision), foresee for Tunisia, a stabilization by 2060 at around 14 million inhabitants, followed by a slight gradual decline (UN 2017). Projections with the constant fertility variant provide quite different results, particularly over longer-term horizons. According to constant fertility variant and starting from comparable values than that reported by INS (2017), the last UN projections (the 2017 revision) expects continued increase: population is expected to reach 14 million by 2040 and almost 18 million by 2100.

The population forecasts will have considerable impact on the drinking water demand. The allocations for drinking water amount to 540 million m^3 in 2010 (about 52 m^3/year/capita) which remain moderate in comparison with per capita demand in developed countries. There is reason to believe that the per capita demand would remain fairly moderate especially as drinking water supply will now rely increasingly on desalination. All these elements together make it rather likely that per capita demand of drinking water would be in the neighborhood of 60 m^3/year/capita by

Table 8.7 Recent updates of the UN population projections for Tunisia

World population prospects	Medium variant		Low variant		High variant		Constant fertility	
	2050	2100	2050	2100	2050	2100	2050	2100
The 2010 revision[a]	12,649	10,891	10,981	6171	14,438	17,730	13,833	14,243
The 2015 revision[b]	13,476	12,494	12,040	7739	14,978	19,020	14,272	16,167
The 2017 revision[c]	13,884	13,321	12,423	8329	15,416	20,112	14,803	17,938

[a]UN (2011), [b]UN (2015), [c]UN (2017), population in thousands

2050. Adopting medium variant, the 2017 revision of population projections indicates that, the population of Tunisia will still below 14 million inhabitants by 2050. The drinking water needs would then amount to 830 million m^3/year. As population is expected to decline for longer term horizons, the drinking water demand would be slowed down just below 800 million m^3/year by 2100. Forecasts with the constant fertility hypothesis, which extends the current demographic state, give higher figures: 890 million m^3/year by 2050 and 1080 million m^3/year by 2100.

However, the recent changes that have occurred in the demography observed in Algeria and Tunisia are rather unique among developing countries where fertility is maintaining elsewhere its downward path. This particular phenomenon is not necessarily permanent (Ouadah-Bedidi et al. 2012) and we adopt thereafter the UN medium assumption on the understanding that the latest updates (e.g. the 2017 revision) integrate the trends observed during the last years.

The future blue water balance of Tunisia

The direct demand (communities, industry, tourism) recorded a significant growth, particularly during the last decade. In the future, competition for water will be more intense, and it seems that the trends in control of use and improvement of their efficiency will be accentuated. The various regulatory, institutional, economic and technical tools will be strengthened to ensure the preservation of the resource. We thus assume that the efficiency of transport and distribution public networks, the estimated values of which for 1970 and 2010 were respectively 60 and 72%, would rise further to 85% by 2050. The agricultural allocations is involved in the water balance with a conversion factor (k) reflecting the overall efficiency at the country level. The conversion factor, the values of which were 52 and 58% respectively in 1970 and 2010 could arise to 75% for 2050.

According to the medium variant hypothesis, the drinking water needs would then amount to 830 million m^3, an increase of 290 million m^3 since 2010. As for industrial uses, it is assumed that by 2050, abstractions will double. Similar assumptions are adopted for direct environmental allocations (conservation of wetlands) estimated at 100 million m^3/year in 2010. It is interesting to note that a significant portion of the additional drinking water resource would be produced by seawater desalination. Recall that with the only completion of the desalination program of SONEDE and its projected capacity expansion, the desalination capacity is expected to reach more than 200 million m^3. This corresponds to a desalinated volume of 15 m^3/year/capita, comparable to the Algerian value for 2012 (FAO 2012). If we assume that this rate could be retained also for industrial uses, the desalination capacity already earmarked for existing programs is expected to exceed 250 million m^3, almost in its entirety from seawater.

As regards treated waste water reuse, the recycling rate remained very low, far below the expected values. Moreover, it does not appear even at the long-term that the expected outcomes are on track to be met especially as projects with structural impact for treated wastewater water reuse and capacity-building (treated wastewater transfers, artificial groundwater recharge) are slowing for objective reasons, related to technical, economic and environmental issues. Consequently, we expect that the total volume of water recycled and reused will remain relatively modest, at 100 million m^3/year by 2050, the double its 2010 value. These estimates, assumptions and projections, as well as other elements related to the BW potential make it possible to establish the future BW balance of Tunisia (Table 8.8).

8.3.3 Outlook for Green Water Resources

Definition and delimitation of green water resources

The National Rainwater Resource is the volume of precipitation measured in an average hydrological year on the national territory. This represents an average volume of 36 km^3/year, with a maximum of 64 km^3/year (1969–70) and a minimum of 21 $km^{3/}$year (1946–47). Delimited to the agricultural area, the Green Water Resource is the part of precipitation which does not runoff nor reach aquifers. It is temporarily stored in the soil and feeds the "Evapotranspiration", either for transpiration and plants consumption, which together form the "productive GW", and for direct evaporation from the soil, which is the "non productive GW". The management of GW consists of maximizing the productive part and minimizing the share of pure evaporation. By "evapotranspiration", we mean in the following the total GW: productive and non-productive.

The linear regression model developed in Chap. 7 provides a rapid and hydrologically coherent means of estimating productive GW, the useful one, referred to cultivated rainfed areas; it is the share of GW that provides regular agricultural production. The analysis of the agrifood water balance highlighted the potentially important role of GW in the existing balance between food supply and demand, and in the equilibrium of food trade balance. GWF of agricultural production from rain-

Table 8.8 Outlook for blue water resources, in million m^3/year

Year	1970	2010	2030	2050	2100
Population (million)	5.1	10.5	12.8	13.9	13.3
Mobilized resource (million m^3)	1065	3960	4268	4575	4575
Among which surface water	450	1940	2320	2700	2700
Among which regulated surface water	345	1340	1670	2000	2000
Among which exploited groundwater[a]	615	2020	1948	1875	1875
Exploitable resource[b]	960	3360	3618	3875	3875
Desalination	0	40	145	250	250
Reuse of treated waste water	10	50	75	100	100
Evaporation—leaks from reservoirs	20	150	175	200	200
Allocated resource[c]	950	3300	3663	4025	4025
Exploited resource[d]	950	3020	3523	4025	4025
Among which irrigation	820	2320	2551	2845	2877
Among which drinking water supply	100	540	717	830	798
Among which non connected industry	30	60	105	150	150
Among which humid zones	0	100	150	200	200
Used and consumed resource	516	1894	2514	3190	3186
Among which irrigation	426	1346	1696	2134	2158
Among which drinking water supply	60	389	563	706	678
Among which non connected industry	30	60	105	150	150
Among which humid zones	0	100	150	200	200
Efficiency of irrigation systems	52%	58%	66%	75%	75%
Efficiency of drinking water systems	60%	72%	79%	85%	85%

[a]Groundwater with a salinity of less than 4 g/L; and overexploitation issues resolved
[b]Regularized surface water + long-term exploitable groundwater
[c]All allocations are used in 2050
[d]In 2010, part (20%) of surface water allocated to irrigated public zones is not used

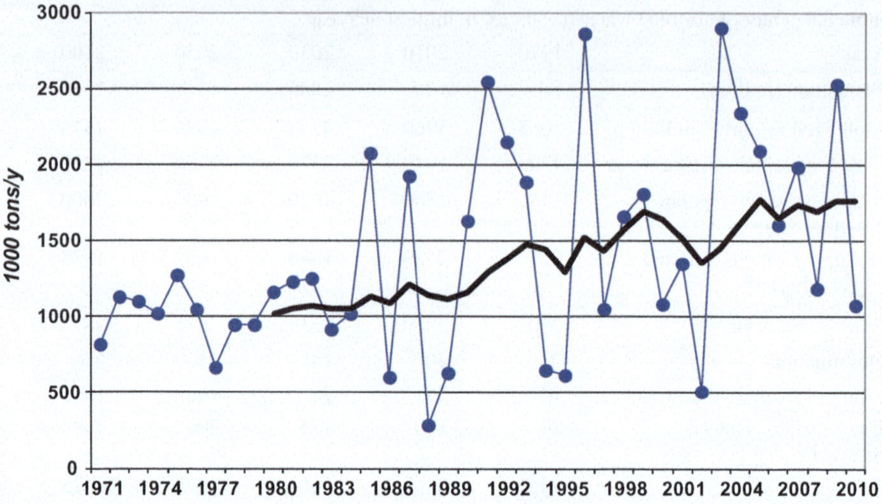

Fig. 8.2 The yearly cereals production progress in Tunisia (1000 tons/year), and the 10 years moving average curve (data from Ministry of Agriculture databases)

fed and irrigated crops represents an important amount of water resources far larger than the volume of the BW effectively used in food production. Moreover, while the total area cultivated under rainfed conditions remained almost stable, the WF of rainfed production has gradually increased. This means that the significant progress in productive GW is due mainly to the improvements of rainfed crop yields.

Figure 8.2 shows for the example of cereals, the important production increase which has grown by a factor 1.8 during the last four decades. Sown areas remaining stable, that is due to remarkable yields progress. This applies also to other rainfed crops (fruits, vegetables, grasslands products) for which substantial increases have been recorded.

Irrigation development becomes now limited, partly because of the high level of water resources mobilization or over-exploitation as is currently the case for a large number of aquifers trough the country, and partly because of water demand increase by other sectors (collectivities, industry, tourism) even if the favorable outlook for seawater desalination development could alleviate pressure on surface and groundwater resources. In return, rainfed agriculture has a great potential for GW development: it is likely that the observed past yields progress trends should continue in the future.

However, there are not yet precise studies on long-term development of GW potential. The elements contained in the Water 2050 report (ITES 2011b) indicate that there are relatively large margins of improvement for rainfed agricultural yields, particularly in arable crops and olive cultures. Essentially, the question that we have to settle is the following one: What is the potential for GW resource? And to what point level can we go in the development of rainfed agriculture and rangelands to convert a growing share of non-productive GW into food production?

The «National Green Water Resource (NGWR)»:

Let us remember that according to previously defined notions and concepts, GW is involved in the biomass development, whether of natural (forests, grassland) or agricultural origin (irrigated and rainfed agriculture). We also have the fact that GW comprises a productive part which represents the GW used for agrifood production and the non-productive part which represents essentially direct evaporation. Optimal management of GW should have as objective to enhance the productive part and to minimize the share of pure evaporation. As we consider here water resources directly involved in food production, discussions on GW involved in forest biomass production, is omitted. The analysis is thus focused on GWF related to agricultural and pastoral activities. First, let us leave rangelands and irrigated zones, and consider only rainfed cultivated areas. The NGWR is then defined by actual evapotranspiration on the rainfed cultivated areas. We can consider constant surfaces, given their insignificant change. According to different studies led in Tunisia, GW from rainfed agriculture represents in first analysis between 6.7 and 12.4 km^3/year. (i) the first value (6.7 km^3) is the GW related to rainfed crops production estimated using the linear regression model previously adjusted on the basis of national data over the period 2001–2010 (see §7.2). Calculated by applying the useful average rainfall depth to cultivated area, this estimation represents the productive GW (or crops production GWF) in average hydrological year (Table 7.5); (ii) the second value (12.4 km^3) comes from the Evapotranspiration estimation during a relatively wet hydrological year (2006–2007), using remote sensing analysis. The latter value represents both productive and non productive GW resource over the whole cultivated rainfed areas. Beyond their demonstrative and pioneer character, results of actual ET modelling using satellite images of cultivated areas (see §4.3.3 and Waterwatch 2008) have been the starting point to make an evaluation of GW resource at national scale (Table 4.5). The ET maps, generated for two hydrological years (2000–2001 and 2006–2007) still constitute the only markers allowing for a framing of the ET temporal variability. Let us recall the main elements on the rainfed areas:

2000–2001—Rainfall supply: 8.9 km^3; Evapotranspiration: 9.6 km^3
2006–2007—Rainfall supply: 14.7 km^3; Evapotranspiration: 12.4 km^3

With these two years, represented by two points, we build a linear empirical relation between Rainfall and ET, i.e. GW. This model, although very rustic and fragile, will generate a plausible series of ET, comparable to that of precipitation (Fig. 8.3), with some deficit years, like 2000–2001, where Evapotranspiration must draw on the precedent year soil water reserve. Such a situation of extreme water stress is exceptional, and it can be noted that on the historical series of almost 100 years, there has never been observed a pair of two consecutive deficit years. The series characteristics are summarized in Table 8.9, which establishes the total rainfed GW at 11.6 km^3 in average hydrological year.

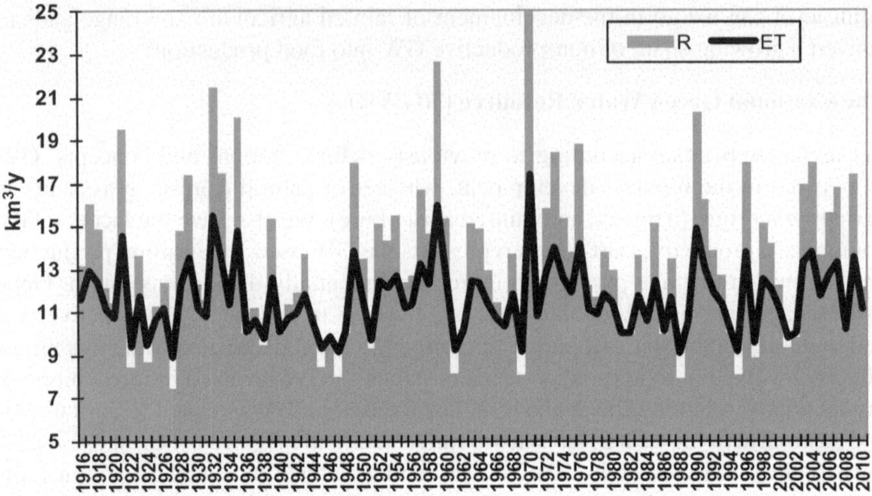

Fig. 8.3 Rainfall (R) and Evapotranspiration (ET, green water) series on rainfed cultivated areas, km^3/year from 1916 to 2010

Table 8.9 Characteristics of 1916–2010 statistical series

	Rainwater resource km^3/year	Evapotranspiration km^3/year
Mean	13.3	11.6
Minimum	7.5	8.9
Maximum	25.5	17.4
Standard deviation	3.6	1.7

The difference resulting from the comparison of the averaged values provided by the two assessments of GW-on cultivated rainfed areas-at the national scale would merit further discussion. The average actual ET carried out using remote sensing modelling represents the ultimate potential that could be extracted by plants from the soil, or the national GW resource, estimated at 11.6 km^3/year. Only part of this potential is productive and could be converted into food production. As the linear regression model relates the GW resources to crop production, it provides an evaluation of the average productive part of the GW actually used by the cultivated rainfed agriculture, estimated at 6.7 km^3/year (see Table 8.5). As compared to the actual ET assessment, the share of GW involved in food production (or crops production WF) represents less than 60% of the national GW resources. Part of the difference is due to the fact that the arable land available for rainfed crops is not cultivated in totality and part of the cultivated area is not harvested due to possible failure of their crops. A last part of this difference is due to the fact that evapotranspiration does not occur throughout the year, especially for annual crops such as cereals whose active period is limited in time.

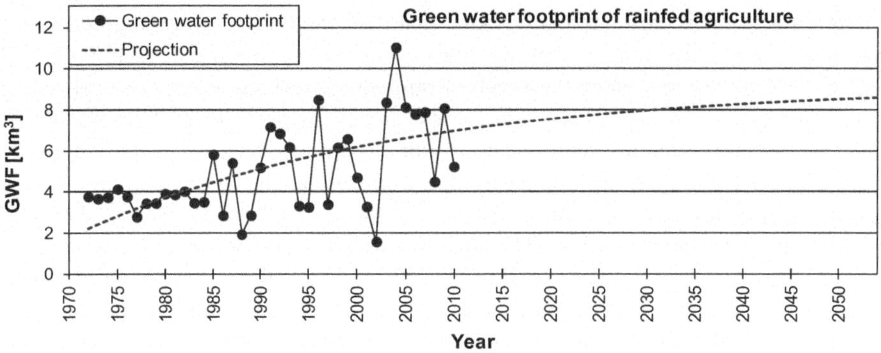

Fig. 8.4 Forecast of green water foot print of rainfed crop production

The unused part of GW represents the margin that could be theoretically used to enhance rainfed crops production. This represents a potential of 4.9 km^3/year. If one assumes that, at the long term, the half of this resource could be reasonably converted into crop production, the WF of rainfed crop production could reach 9.2 km^3 which can be thought of as an ultimate upper limit. Building on the assumption that the growth of the GWF would be proportional to the gap between the current GWF and the upper limit, an exponential model adjusted on the previous series 1972–2010, provides the forecast of Fig. 8.4. According to this projection, the GWF of rainfed arable lands increases to 8 km^3 by 2030 (for the average hydrological year) and 8.5 km^3 by 2050. It would reach 9.1 km^3 (almost the upper limit) by 2100.

The second component of GWF of crop production is related to irrigated crops. The total WF of irrigated agriculture is estimated at 2.3 km^3, of which the share of GW amounts to 0.9 km^3, (Table 8.5). With regard to irrigation, we have seen (Table 8.8) that, by a better utilization of mobilized surface water, allocated volumes would increase by 22%, but with a clear improvement in the irrigation efficiency, the increase in water consumption by crops could increase by 60% over the long term, from 1.35 to 2.16 km^3/year. This will be accompanied, on average, by a corresponding increase in irrigated areas, and thus a comparable increase in mobilized GW, from 0.9 to 1.45 km^3/year. Assuming that GW surpluses on irrigated crops are to be drawn from GWF of rainfed area, the GWF potential of crop production would then be around 10 km^3/year. This represents 80% of actual ET on arable land estimated at 12.5 km^3.

To this resource should be added the GW generated by precipitation on rangelands. There is no sufficient data that permit a formal assessment of GW resource from rangeland. However we might seek to set some benchmarks from previous developments and outcomes. We have on the one hand the average area-weighted rainfall on rangelands calculated in Chap. 4 (see §4.7.2). It indicates that rangelands, large areas of which are located in the central and southern parts of the country, receive less rainfall (almost the half) than cultivated areas located at the wettest areas of the country. As rangelands and arable lands occupy equivalent total areas, actual

ET on rangelands is estimated at 6.6 km^3 just over the half of average green water resources (productive and non productive) on cultivated areas (see Table 4.8 and Fig. 4.18). On the other hand, the GWF of rangeland production has been estimated at 1.3 km^3 (see Table 8.5). This value was obtained by comparing the total WF of national animal production with the WF of all animal feeding (see §7.4). Assuming that rangelands productivity improvements would be at the long term comparable to the one for rainfed agriculture (40% more), the GW related to extensive animal production would be 1.8 km^3 by 2100 leaving a reserve of 4.8 km^3.

According to these figures, the National Green Water Resource potential would amount on average to 19.1 km^3/year whose productive part is estimated at 11.8 km^3/year. Recall that this potential does not consider GW supporting biomass production on forests. The total GW (Evapotranspiration) on forests has been estimated in Chap. 4 at 4.5 km^3 (see §4.7.2 and Table 4.8). Taking into account the ET from forests and from irrigation, we retrieve the amount of the total ET estimated in Chap. 4 at 25 km^3.

8.3.4 Outlook for Food Demand

Impact of dietary changes on water requirements for food demand

The water balance analysis of food needs carried out in Chap. 7, makes it possible to assess the various elements of the food demand and their change. It appears that while the Tunisian population has slightly more than doubled over the past four decades, water requirements for food demand have more than tripled. This means that the per capita food demand rose by more than half in that time period, from about 1000 m^3/year/capita in 1970 to nearly 1600 m^3/year/capita in 2010.

The significant increase of water requirements of the food demand indicates that Tunisia is still in a dietary transition phase. Apparently, this is also the case for other Mediterranean countries, like Turkey and Morocco for example. It is appropriate in this regard to observe that diet transition may be more or less rapid. Figure 8.5 shows the evolution of meat consumption in a number of representative Mediterranean countries and in the USA. Within five decades, the total meat consumption increased by a factor 4 in Portugal and by 2 in Turkey, Morocco and Tunisia.

In comparison with countries that have already achieved their diet transition (North American and European countries), the consumption of meat in Morocco and Tunisia remains relatively low; three times lower than in France and Portugal and four times lower than in USA in 2010. In return Morocco and Tunisia have consumption levels of foodstuffs of crop origin comparable and even higher than those of European countries with however differences of internal distribution: high level of cereals (twice than European countries) and vegetables, (Fig. 8.6). Moreover, with the increase over the last decades, the consumption levels of foodstuffs of crop origin are on track to reach the maximum values recorded for Turkey. These stabilized since the 1990s.

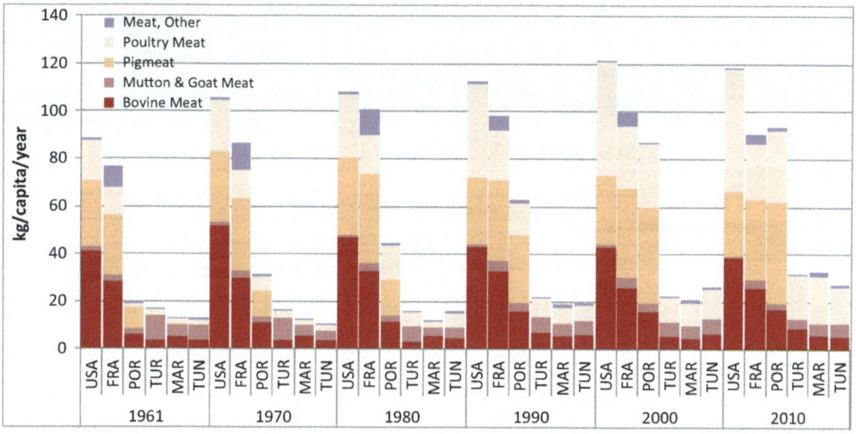

Fig. 8.5 Change, from 1961 to 2010, in total meat consumption of some Mediterranean countries (France FRA, Portugal POR, Turkey TUR, Morocco MOR, Tunisia TUN) and the USA (kg/capita/year; data from FAO 2017)

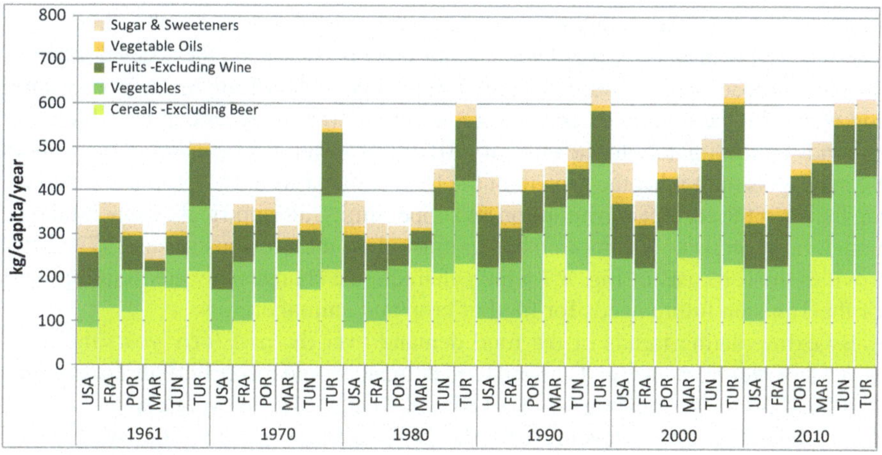

Fig. 8.6 Food demand of crops origin in Mediterranean countries and the US (kg/capita/year; data from FAO 2017)

Water requirements for per capita food demand

One may consider that the Tunisian consumption of foodstuffs of crops origin, which reached WF levels similar to those in Turkey (1100–1200 m^3/year/capita), is on the way towards stabilization. On the other hand, the total meat consumption of European countries (France and Portugal) has already stabilized and started a slower decline phase. It amounts at around three times the total meat consumption in Turkey, Morocco and Tunisia. Absent in the dietary of these three countries, the pig meat

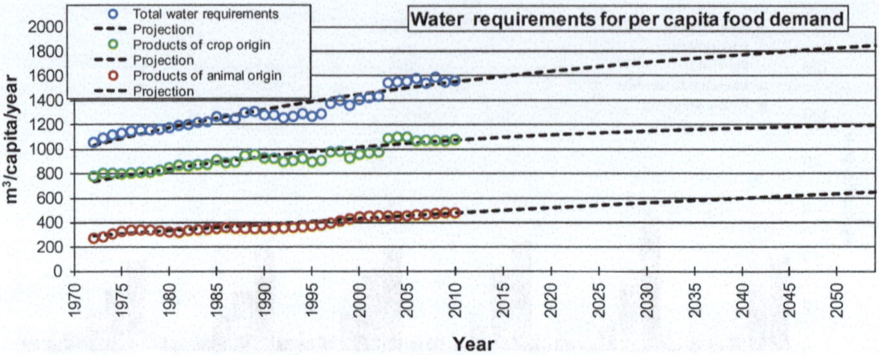

Fig. 8.7 Outlook for food demand water requirements

represents an important part of the European meat consumption: a full third in France and almost a half in Portugal.

Given the structure of the Tunisian diet which remains relatively low in red meat products, one may expect an increase in consumption of foodstuffs of animal origin accompanied with an increase of water requirements. To predict these developments, one can suppose, that the ultimate upper limits of per capita Tunisian meat consumption would be the double its actual value (around 1000 m^3/year/capita) which corresponds approximately to the current (2010) European meat consumption without pig meat. Assuming that the growth of the food demand would be proportional to the gap between the current consumption and upper limits, an empirical exponential type model, built on these assumptions and adjusted on the previous series 1972–2010, provides the forecasts of Fig. 8.7 which characterize changes in water requirements for the Tunisian future needs for food of crop and animal origins.

By extrapolating trends of the food demand over the past forty years,the long-term water requirements for Tunisian food demand would reach 1710 m^3/year/capita in 2030,1830 m^3/year/capita by 2050 (Fig. 8.7) and 2020 m^3/year/capita by 2100. Compared to the present level (2010), the Tunisian meat consumption will grow by 17% at the horizon 2030, 32% at the horizon 2050 and by 64% at the horizon 2100. Despite this significant increase in meat consumption, the contribution from animal products in diet composition will remain still relatively low, below 40%, increasing from 31% in 2010 to 35% in 2050 and 39% in 2100. Tunisia will keep with the basic characteristics of Mediterranean diet pattern which, even over long time periods, will remain far from the European one, for which around two thirds of the water requirements for food demand are related to food products of animal origin.

8.3.5 *Outlook for the National Food Water Balance*

National water balance and water dependency index evolution

The outlooks for BW and GW developments make it possible to predict the future food production WF. The comparison of these results with water requirements for food demand allows estimating the contribution of VW and determine the Water Dependency Index. The results are presented in Table 8.10.

The outlook for BW relies on substantial increase of seawater desalination and on stronger control of all water uses to boost their efficiencies. The implications of such programs on water security are obvious: the BWF of the irrigated crop productions will rise from 1.4 km^3 in 2010 to 2.1 km^3 by 2050 (an increase of 50%). At the same horizon, GW involved in food production is expected to increase by almost 25%. However, despite the expected progress of both irrigated and rainfed agriculture, the population growth and increased food demand will lead to a gradual increase of the VW and correlatively to an increase of the water dependency index. Recall that these results are carried out assuming constant agricultural rainfed area. We may expect that it is possible to move forward in improving the national water balance if appropriate measures can be taken to go further in developing GW resource by increasing cultivated rainfed area.

Demographic change underlying the related projections is proving to be very important factor in that it determines the future needs. According to the constant fertility assumption, the population will be larger in the future and the water requirements of the food demand will correspondingly increase. While assumptions about demography have not too much effect on the projection results at the horizon 2030, and to a lesser extent at the horizon 2050, their effects at longer term (horizon 2100) are however considerable. The medium variant expects a population decline during the last half of the century which would bring back the population below its 2050 level, a population of 13.3 million inhabitants according to the 2017 revision (UN 2017). Taking account the excepted increase of the per capita water requirements of the food demand (2020 m^3/year/capita in 2100), the VW balance deficit would be around 12.9 km^3, not much higher than the forecasted value for 2050 (12.3 km^3). The same goes for the Water Dependency Index the value of which would be around 48–49%. With the constant fertility assumption the population in 2100 would be much higher and the water deficit will amplify considerably reaching 22.4 km^3. The Water Dependency Index would be around 62%.

Nevertheless, promoting rainfed agriculture which is highly exposed to climate hazards requires the development of appropriate strategies to provide solutions to specific problems, especially those associated with production variability. It has been shown (see §7.2.6) that the variability of GW resources associated with rainfed crops production is roughly two times higher than that of the rainfall. The results carried out revealed that, one in five years (frequency 20%), the GWF of food production, could rise (or fall) by some 35%. This rate of change could reach 55% one of every ten years (frequency 10%). Taking into account the annual variability of GW, Table 8.11.

Table 8.10 Outlook for average food water balance

Year		1970	2010	2030		2050		2100	
				MV[a]	CF[b]	MV	CF	MV	CF
Population (million)		5.1	10.5	12.8	13.1	13.9	14.8	13.3	17.9
Blue water footprint (km^3)		0.4	1.4	1.7	1.7	2.1	2.1	2.2	2
Green water footprint (km^3)	Rainfed crops	3.6	6.7	7.8	7.8	8	8	8.5	8.6
	Irrigated crops	0.3	0.9	1.1	1.1	1.4	1.4	1.5	1.4
	Rangeland	0.5	1.3	1.5	1.5	1.6	1.6	1.8	1.8
Water requirements for food needs	m^3/year/capita	1000	1600	1710	1710	1830	1830	2020	2020
	km^3/year	5.1	16.8	21.9	22.4	25.4	27.1	26.9	36.2
Virtual water balance deficit	m^3/year/capita	59	619	765	786	888	945	967	1249
	km^3/year	0.3	6.5	9.8	10.3	12.3	14.0	12.9	22.4
Water dependency index (WDI)		6%	39%	45%	46%	49%	52%	48%	62%

[a]Medium variant; [b]Constant fertility

Table 8.11 Impact of climate fluctuations on water dependency index for 2050

Outlook for 2050[a]	Average values	Frequency 20% Min–Max		Frequency 10% Min–Max	
Green water footprint (km^3)	11	7.2	14.9	5.0	17.1
Virtual water balance deficit (km^3)	12.3	16.2	8.5	18.4	6.3
Water dependency index (WDI)	49%	64%	33%	72%	25%

[a]With UN population projection according the medium variant

shows the impact of climate fluctuations on the VW deficit and consequently on the water dependency index.

Table 8.11 shows that GW variability has major implications on the water food balance reflected by significant variations in the water dependency index, the average value of which would be 49% by 2050. It appears that at this horizon the water dependency index would fluctuate considerably: one year in five (frequency 20%) its value could rise to around 64% in dry year or fall to 33% in wet year; and one year in ten (frequency 10%) the water dependency index could reach 72% in dry year or fall to 25% in wet year. However, these results should be put into perspective by stating that they are issued from an analysis of variability over the first decade of the 2000s. Over this relatively short period, the average productivity of rainfed crops is assumed to have low variation. GWF variability deserves further investigations by studies on longer series. These studies must also consider the drift associated with changes in rainfed crop yields.

The yields growth and the future of water footprint

As has been mentioned previously, the linear regression model should be implemented using data carried out over relatively short periods of time so that one may consider that average crop yields and water use efficiency remain relatively stable. The model defines a reference situation at reference period and reference domain, a kind of benchmark model of the productive GW efficiency use. Extending the model over different periods of time for retrospective and prospective analyses can be construed as creating performance comparisons of productive water use efficiency in relation to agricultural productivity.

In reality, the specific consumption of water (per unit of production) of crops is constant only when crop yields are high. Figure 8.8 shows the evolution of the specific water consumption (Evapotranspiration) of cereal crops under different conditions. It indicates that, starting from yield threshold between 3–4 t/ha, the specific consumption is constant (around 1 m^3/kg of grains). Below this threshold, the crop specific consumption could be considered as inversely proportional to the crops yield,

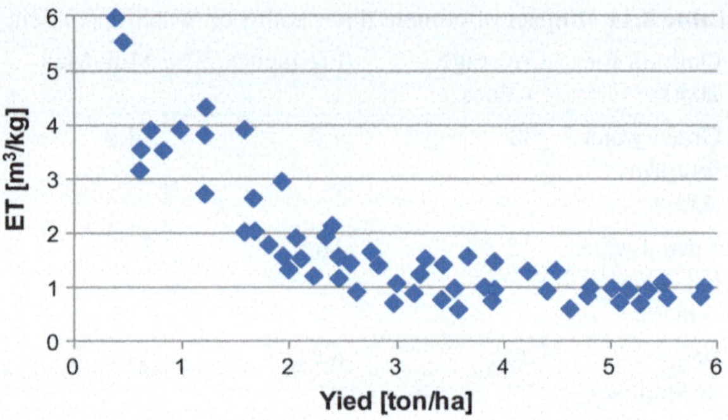

Fig. 8.8 Evolution of the specific water consumption (Evapotranspiration) of cereal crops as a function of their yields. Adapted from Karlberg et al. (2008)

at least for a large part of data. To explain these observations, many authors (Karlberg et al. 2008; Molden et al. 2010; Zimmer 2013) mention the role of vegetation cover, which, depending on its density and degree of development will in a more or less important way, limit direct evaporation and non-productive transpiration.

Cereal rainfed crops in Tunisia have yields of around 1.25 tons/ha in 2010. As the cereals crop yields are relatively low, the improvement in crop yields also improves the productivity of water and vice versa. This may explain the important progress in cereals productivity observed during the last four decades: cereals crop yields have recorded steady progress from nearly 0.6 t/ha in 1971 to 1.25 t/ha in 2010. Assuming that the specific consumption becomes constant equal to 1 m^3/kg when crop yields reach 2.5 t/ha, and noting that in 2010 the specific consumption is 2 m^3/kg for an average yield of 1.25 t/ha, a proportionality hypothesis leads to estimate the possible specific consumption values at different periods: slightly higher in 1970, lower in the future if one supposes that yields continue to increase. Expected levels of gain in productivity of cereal crops are nevertheless to be relativized when we consider increase of the whole rainfed crops, knowing that olive tree yields have increased only about 15% in the last 40 years. Finally, the surplus production generated by productivity increases results in equivalent decreases in foodstuffs imports with positive effects on the WDI. A rough estimate makes it possible to predict the future decrease on the estimated value of WDI (Table 8.10) at around 2 and 5%, respectively by 2030 and 2100 for the medium variant demographic hypothesis. These differences, although very important, remain in the order of magnitude of the margins of uncertainty in the main parameters of future WDIs, namely the population and the WF.

How to optimize virtual water in the future

Food and water balance of Tunisia will remain in deficit and net Virtual Water (VW) flows will participate in a structural manner to national food security. In general, Tunisia has interest to import the maximum of high water consuming agricultural products to compensate its internal deficits while exporting agricultural products with low water consumption and high added value. This orientation principle makes it possible to reduce the level of exploitation and degradation of internal resources while meeting the water and trade deficit. In this context, the only constraint is the

Table 8.12 Change in average rainfall in the second half of this century compared to the second half of the twentieth century (according to de Marsily 2006)

Geographical region	December to March	June to September
Equatorial Africa	[+25%]	[+10%]
Sahelian Africa	Uncertain	[+30%]
North Africa	[−15%]	[−10%]
South of Europe	Uncertain	[−20%]
France	[+15%]	Uncertain
Scandinavia	[+25%]	[+15%]

excessive import of strategic commodities, those that constitute the basis of Tunisian food, particularly grains and edible oils. It is essential that a country can guarantee some independence in the production of staple foods to ensure at all times a minimum food security. But it is also important in this context to define the quantities of strategic food it is essential to locally produce at affordable prices. Indeed, all quantities of cereals produced or imported are not intended actually to basic supply and waste occurs at various levels of the value chain. Furthermore, and as for the management of water demand, it is essential to design and implement a policy of food demand management, which aims to rationalize demand foodstuffs whose production is still closely linked the issue of water (Hamdane 2013).

8.3.6 The Impact of Climate Change on the National Water Security

At the global level, Table 8.12 shows the precipitation changes calculated for a likely scenario by the Intergovernmental Panel for Climate Change (IPCC). Results from the latest IPCC reports (IPCC 2008, 2009; Collins et al. 2013), based on an average of the different models, are consistent with these predictions.

Despite uncertainties, the major consequences provided for the distribution of water resources in the Mediterranean region are:

(i) a sharp decline, on average, of the soil moisture content (higher Evapotranspiration due to temperature increase and rainfall decrease, particularly in summer) with a consequent drop in rainfed agriculture production;
(ii) an increased risk of agricultural drought, which occurs in spring and summer;
(iii) a risk of stronger hydrological droughts occurring in autumn and winter, influencing surface flows and aquifers recharge.

Note that if the uncertainties of the models used for simulations are unevenly distributed around the globe, most climate projections converge on a larger drying of the Mediterranean climate.

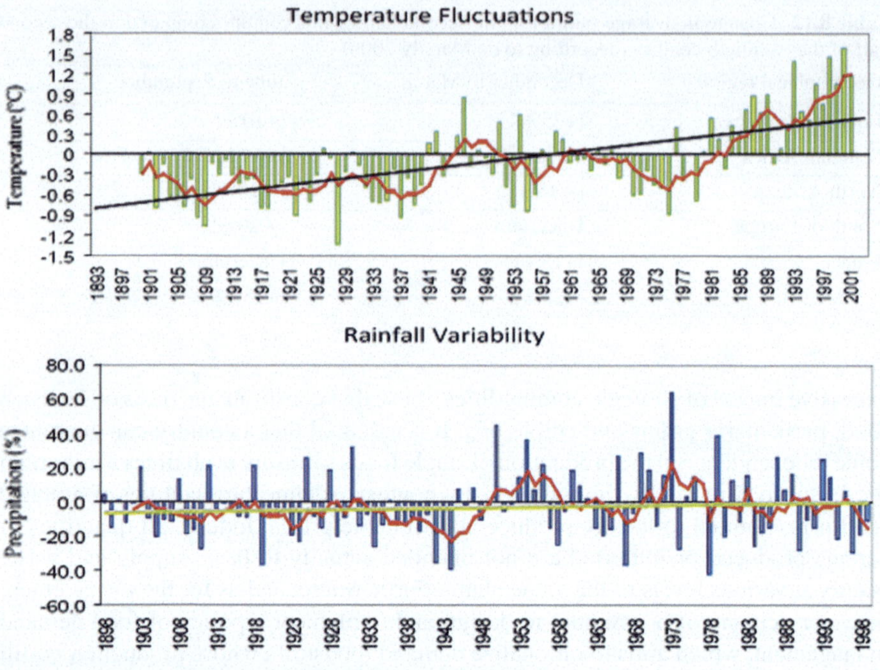

Fig. 8.9 Changes of the mean temperature and precipitation during the 20th century in Tunisia. *Source* MARH et al. (2007)

Scenarios for Tunisia:

Regarding particularly Tunisia (MARH et al. 2007), the analysis of changes in mean temperature and rainfall over the past century (Fig. 8.9) shows (King et al. 2007):

(i) For temperature, the main observation is a significant increase of 1.2 °C over the past century. This increase is higher than the global average increase of 0.7 °C indicated by the 2001 IPCC report.
(ii) For rainfall, no significant trend is detected. Note however that the reference period 1961–1990 is characterized by the highest variability (highest standard deviation) in comparison with previous periods 1931–1960 and 1901–1930.

In the context of climate change, the future evolution of rainfall and temperature could be modeled on the global scale but with relatively large uncertainties. The extent and exact nature of future changes remain to be specified, so care must be taken to implement comprehensive information to regional and local scales. The chosen medium scenario confirms these difficulties: if the general increase in temperatures is admitted by all, the evolution of rain is sometimes positive and sometimes negative, variable according to regions and seasons, and the resulting decline is very low, which will not significantly affect the surface flows and contributions to dams (Lahache Gafrej 2007).

But the decrease in summer precipitation and rising temperatures and PET will increase the soil water deficit. This will create additional difficulties for rainfed agriculture, and even greater exploitation of groundwater. Furthermore, the current sea level rise can exacerbate certain coastal groundwater salinization and reduce the long-term potential groundwater resources. Climate changes result for Tunisia by increased climate variability. In particular, extreme events (droughts, floods) are expected to increase in frequency and intensity, and very dry years to occur more often in the future.

Climate change and the cereals supply evolution:

Agriculture is particularly affected by climate change, in particular rainfed agriculture widely involved worldwide in cereals production. Tunisia, which wheat supplies depend for more than the half on imports, is affected by the climatic issue not only for its impacts on local agricultural production, but also for its incidences on the future of cereals world production, and for the effects that may have on the growing Tunisia's grain bills. Virtual water transfers, which reflect water resources globalization, also translate into a globalization of the climate issue. The abundant literature on climate change impacts on agriculture provides estimations for several climate change scenarios in different regions of the globe. Although the global production appears stable (Parry et al. 2004), crop yields are expected to be differently affected: some regions may improve their agricultural production, in France for example (ONERC 2009) whereas others, such as USA, will suffer from yield losses (Schauberger et al. 2017). The regional differences in crop production are likely to grow leading to polarization effects with expected increases in prices (Parry et al. 2004). Tunisia has to anticipate these changes and put in place adaptation strategies. It must, more generally, equip itself with cognitive ability and prospective tools to measure the impacts of any changes of trade terms on food security: market liberalization, changes in agriculture policies and food subsidies, and thus minimize the long-term predictable deterioration of WDI as a result of climate change.

References

Bachta MS, Zaïbet L (2006) Les innovations institutionnelles comme adaptation à l'évolution des périmètres irrigués en Tunisie, sém. Wademed, Cahors, Cirad

Ben Mammou A, Louati MH (2007) Évolution temporelle de l'envasement des retenues de barrages de Tunisie» Revue des sciences de l'eau. J Water Sci 20(2):201–210

Besbes M, Chahed J, Bargaoui Z, Mattoussi MS, Mhiri A (2002) Water future, a new challenge for Tunisia. Int. Rep. Tunisian Institute for Strategic Studies, Tunis

Besbes M, Chahed J, Hamdane A, de Marsily G (2010) Changing water resources and food supply in arid zones: Tunisia. Water Sustain Arid Reg 2010:103–121

Chahed J, Besbes M, Hamdane A (2007) Stress hydrique et sécurité alimentaire, une vision intégrale des ressources en eau. La Houille Blanche 3:52–57

Collins M, Knutti R, Arblaster J, Dufresne J-L, Fichefet T, Friedlingstein P, Gao X, Gutowski WJ, Johns T, Krinner G, Shongwe M, Tebaldi C, Weaver AJ, Wehner M (2013) Long-term climate change: projections, commitments and irreversibility. In: Working group I contribution to the IPCC fifth assessment report (AR5), climate change 2013: the physical science basis

de Marsily G (2006) Les Eaux Continentales, by Académie des Sciences; coordonné par G. de Marsily, EDP Sciences, Paris

FAO (2012) AQUASTAT: FAO's global water information system, the Land and Water Division. http://www.fao.org/nr/water/aquastat/main/index.stm, consulted on 11 Oct 2017

FAO (2017) FAOSTAT: food and agriculture data. Statistics Division. http://www.fao.org/faostat/en/#data, consulted on 11 Oct 2017

GEORE (2001) GEstion Optimale des Ressources en Eau. Projet GEORE, DGBGTH, Ministère de l'Agriculture

Hamdane A (2013) La triade eau bleue, eau verte, eau virtuelle. Demeter 2013 Paris. Economie et stratégies agricoles

INS (2017) http://www.ins.nat.tn/fr/themes/population, consulted on 23 Sept 2017

IPCC (2008) Climate change and water, technical report VI. Inter governmental panel on climate change

IPCC (2009) The copenhagen diagnosis; Updating the world on the latest climate science, Interim Report Inter governmental panel on climate change. Climate Change Research Centre, 2009

ITES (2011a) Etude stratégique - Eau 2050, Annexe 2, Mobilisation des ressources en eau, *int.rep.* Tunisian Institute for Strategic Studies

ITES (2011b) Etude stratégique - Eau 2050, Annexe 3, réflexion sur l'agriculture pluviale, *int.rep.* Tunisian Institute for Strategic Studies

Karlberg L, Barron J, Rockström J (2008) Water productivity and green water management in agro-ecosystems. In Förare J (ed) Water for food. The Swedish Research Council for Environment, Agricultural Sciences and Spatial Planning, Stockholm, pp 63–77

Khaldi R, Naili A (2001) Dynamique de la consommation de lait et de produits laitiers en Tunisie. Options Méditerranéennes, Sér. B/ n° 32, 2001

King L, Nasr Z, Almohamad H, Maag CC (2007) Le Climat; in «MARH, GTZ, Gopa, Exaconsult: Stratégie nationale d'adaptation de l'agriculture tunisienne et des écosystèmes aux changements climatiques», Ch.7.2; Tunis Janvier 2007

Lahache Gafrej R (2007) Les Ressources en eau, in «MARH, GTZ, Gopa, Exaconsult: Stratégie nationale d'adaptation de l'agriculture tunisienne et des écosystèmes aux changements climatiques», Ch.7.3; Tunis Janvier 2007

Louati MH, Khanfir R, Alouini A, El Euchi M, Marzouk M, Frigui L (1998) Eau21; Stratégie du secteur de l'eau à long terme; Ministère de l'agriculture, Tunis

MARH, GTZ, Gopa and Exaconsult (2007) Stratégie nationale d'adaptation de l'agriculture tunisienne et des écosystèmes aux changements climatiques

Molden D, Oweis T, Steduto P, Bindraban P, Hanjra MA, Kijne J (2010) Improving agricultural water productivity: between optimism and caution. Agric Water Manag 97(4):528–535

National Oil Office (2017) http://www.onh.com.tn/index.php/fr/2016-05-23-14-44-46/la-production, consulted on 11 Oct 2017

Office of cereals (2017) Infrastructures de stockage; in: http://www.oc.agrinet.tn/index.php/fr/presentation/infrastructure-de-stockage, consulted on 11 Oct 2017

OONERC (2009) Climate Change impacts, adaptation and associated costs in France, National observatory on the effects of global warming, document prepared for the side event held on 11 December 2009 United Nations Climate Change Conference (Copenhagen, 7–18 December 2009), http://research.fit.edu/sealevelriselibrary/documents/doc_mgr/419/France_Studies_on_CC_Adaptation_-_AFPCN_2009.pdf, consulted on 11 Oct 2017

Ouadah-Bedidi Z, Vallin J, Bouchoucha I (2012) La fécondité au Maghreb: nouvelle surprise, Population et Sociétés, bulletin mensuel d'information de l'Institut national d'études démographiques, numéro 486, février 2012

Parry ML, Rosenzweig C, Iglesias A, Livermore M, Fischer G (2004) Effects of climate change on global food production under SRES emissions and socio-economic scenarios. Glob Environ Change 14(1):53–67

Population Reference Bureau (2016) 2016 world population data sheet with a special focus on human needs and sustainable resource, http://www.prb.org/pdf16/prb-wpds2016-web-2016.pdf, consulted on 11 Oct 2017

Schauberger B, Archontoulis S, Arneth A, Balkovic J, Ciais P, Deryng D, Pugh TA (2017) Consistent negative response of US crops to high temperatures in observations and crop models. Nat Commun 8:13931

SONEDE (2017) http://www.sonede.com.tn/index.php?id=56, consulted on 11 Oct 2017

UN (2011) World population prospects: the 2010 revision, volume I: comprehensive tables. ST/ESA/SER.A/313. United Nations, Department of Economic and Social Affairs, Population Division

UN (2015) World population prospects: the 2015 revision, key findings and advance tables. Working paper no. ESA/P/WP.241. United Nations, Department of Economic and Social Affairs, Population Division

UN (2017) World population prospects: The 2017 revision, key findings and advance tables. Working paper no. ESA/P/WP/248. United Nations, Department of Economic and Social Affairs, Population Division

WaterWatch (2008) Water balance and evaluation of water saving investments in tunisian agriculture; draft version, 2008. Study under assignment of the World Bank in Washington D.C, USA; Authors: S.J. Zwart and W.G.M. Bastiaanssen from WaterWatch in Wageningen

World Bank (2012) The grain chain, food security and managing wheat imports in Arab countries, Washington, D.C

Zaara M (2008) Le dessalement en Tunisie pour l'amélioration de la qualité de l'eau potable desservie; RMSU MEDA Water Programme, ADIRA Workshop, Marrakech

Zimmer D (2013) L'empreinte eau. Les faces cachées d'une ressource vitale. Charles Léopold Meyer, Paris

Chapter 9
Elements for a Conceptual Model of National Water Security

Contents

9.1 Introduction

The idea of water insecurity and threats to water resources has been recurring for almost 50 years, with the work of the Club of Rome, the MIT Model on The Limits to growth, and awareness of excessive industrialization side effects in Europe and the United States due to massive surface-water pollution. With the sustainable development concept and observations of the first damage underlying the Green Revolution in India, where about 22 million wells today collect 250 km^3/year of groundwater for irrigation, water shortage predictions have flourished, placing arid countries such as Tunisia among the first threatened nations. For 25 years, the experts have therefore been announcing severe water shortages, first by 2000 and then for increasingly distant times. The delay in the arrival of the announced "catastrophe" is probably linked to the following observations: (i) steady decline in population growth estimates in many regions of the world, with some contradictory new trends however; (ii) water management has, on the whole, progressed, despite the demographics but with some economic development and empowerment of the populations; (iii) internal problems of water development and supply have often been, if not solved, at least on the way to solutions. In parallel, the scarcity concept definition has moved from impending water-shortage crises and water wars to the necessity of moderating conflicts and regulating the food trade. Thus, the concept

M. Besbes et al., *National Water Security*,
https://doi.org/10.1007/978-3-319-75499-4_9

of food self-sufficiency has gradually changed into a food-security concern, due to trade globalization. Consequently, water security paradigm have moved from a spatially limited national vision based on water stress considerations to an open vision, which, with the generalization of Virtual Water (VW) and Water Footprint (WF) concepts, leads to water resources globalization. Given the dialectic water food relationship, the approach to managing water resources comes inevitably through a holistic view that considers the multiple stress factors likely to affect water supply.

Developments in the understanding of the water security paradigm are illustrated through the example of Tunisia. Throughout the book, we have proposed to analyze the Tunisian case in order to introduce and illustrate the concept of National Water Security. So far, provisions, methods and tools developed for the Tunisian case, serve as a support for the development of a National Water Security conceptual model which can be duplicated to other arid countries.

9.2 Perspectives on Long-Term Holistic Water Balances

9.2.1 Respective Roles of Blue and Green Water in Food Demand Supply

The formulation of the water balances confirms that much of mobilized water is for agriculture production. Presumably, the increase of direct water use could result in a reduction of agricultural water allocations that would result on the long-term in relative decline of the irrigation sector, especially in some rural regions. Such a situation would constitute an economic and social regression that Tunisia will try to avoid by continuing or adapting its water policy to support food security. The ongoing programs express a desire for development of Blue Water (BW) resources, which in addition to the completion of the mobilization of surface water, provide a real involvement in the production of non conventional resources especially by desalination.

The second level of commitment concerns the continuation of efforts towards mastery of water usage and increasing efficiencies. However, and beyond the scope of the national water project and its financial implications, Tunisia will be called to provide answers to uncertainties that may affect achievement of such a program, associated with its environmental impact, with the future status of the resource and the technical and economic conditions of production and operating unconventional resources. It is also needed to engage the institutional, legal and human conditions able to support these ambitious water programs.

Despite its major socio-economic role, the production of irrigated agriculture Water Footprint accounts, for a small share in BW, 13% of food production and only 8% of the water requirements for food demand. Because of the Mediterranean position of Tunisia, the Green Water (GW) contribution to food production is considerable (about 87%). Both irrigated and rainfed sectors play essential roles in food security; different but complementary roles. Irrigated agriculture has a key role in the national economy. Benefiting from the best land resources and managed by an agricultural population generally very enterprising, irrigated agriculture provides diversified production, exempted from weather hazards, allowing internal and external markets supply by various food products including fresh produce. Rainfed agriculture contributes significantly to basic food security and plays a key role in agro-food balance. Considerable possibilities for improvement of agricultural performance, in both rainfed and irrigated components, remain particularly with respect to increasing the productivity of water and soil resources. Improved agricultural performance also depends on networks development and diversification that ensure crop and products storage, processing, marketing and selling.

9.2.2 National Strategies and Action Programs

Increase water efficiency and ensure irrigation sustainability

Water saving and increased efficiency of all uses, including agriculture, are essential components of the water policy. The irrigation sector is expected to produce more and better with the same available amount of water, and maybe less: projections for the sector aim to increase the contribution of irrigation from 35 to 50% of overall agricultural production value. This perspective of irrigation development will amplify the pressure on water and soil resources and increase the risks for these resources. In particular, citrus and dates that contribute longstanding and continuously to agrifood exports are facing many difficulties threatening their sustainability. With 11% in value of agricultural exports and 10% of the irrigated area of the country, these two cultures consume more than 25% of water volumes allocated to the irrigation sector. But a number of problems may make these crops unsustainable as well as the associated trade.

Citrus currently occupy irrigated areas facing chronic groundwater overexploitation in the Cap Bon region, despite costly safeguard conducted through northern waters transfer. As for the date palms, they develop in desert oases using large quantities of slightly renewable water resources, and signs of overexploitation already appear in many areas, due to unplanned extensions and overstated water uses. The sustainability of these cropping systems would be questioned if improvement in the management of available water is not engaged by reducing extensions of cultures and unit consumption of water, and also by focusing on optimal economic returns on productions.

Promote development and use of alternative resources

In the water equation of Tunisia, unconventional resources have occupied a minor role: the reuse of treated wastewater has remained very modest, well below displayed projections, though progress in controlling management and reuse of wastewater will become a necessity in the future, especially since it is available throughout the territory and could provide local water supply solutions. As regards desalination, programs have so far not been of a great ambition. Because of its high cost compared to conventional resources, desalination has always been considered as a solution restricted to situations where water quality deficiency to cover basic needs (drinking water supply, tourism, industrial use) has been observed.

Do we live a turning point regarding the role of desalination in water balance? This can be seen in the light of SONEDE's new and ambitious sea water desalination projects to secure drinking water supply to the southern coastal regions of Sfax, Gabes and Jerba. If economically, desalination is able to compete with conventional modes of transferred water resources, this should have positive consequences: in addition to relief pressure on resources, this fresh water intake can be reused. Thus, it is likely that desalination is expected to grow. Nevertheless, it is important to note in this regard that Tunisia will face two shortages simultaneously in the future: water and energy, despite its situation in a favorable environment to the production of alternative energy. The future of desalination will therefore depend on an integrated vision on the water question on the one hand, and the energy question on the other.

Develop a comprehensive strategy to improve green water use efficiency

Prospective forecasts indicate that regardless of the assumptions on the development and conservation of conventional and unconventional resources, the structural deficit in food demand and the corresponding water balance will increase if strong measures are not taken to reconsider the role of rainfed agriculture. The latter, which is a key factor in development of water and soil resources, has not experienced the same boom that irrigated agriculture and has not yet found the place it deserves in terms of technological improvement and in terms of intellectual and financial investments.

The principle of maximum valorization of m^3 of water, relied on by the Water Code, should first be expanded to Green Water (GW) mobilized by rainfed agriculture which will lay the legal foundations of this sector development. In practical terms, the promotion of this sector requires that we work toward advancing, among others, rehabilitation of traditional arid systems of water development and soil resources in areas that are suitable. These concern rainwater collection methods, water spreading and conservation in the soil which can significantly increase agricultural production. Scientifically, it is necessary that we develop knowledge on the physical, physiological and genetic mechanisms of resistance of crops to drought, to increase storage capacity and water use. Far from being outdated, systems and techniques of crops adapted to aridity are still extremely topical issues.

The promotion of rainfed agriculture also requires that we take measures both structural and non-structural to better manage its production variability. Structural measures will concern interventions and infrastructure construction (roads, grain silos, food reserves) designed to implement an effective strategy to promote food production, combating crop production losses, handling and storage. Non-structural steps to be taken refer to a range of accompanying actions to help farmers developing their capacities. In particular, appropriate drought management, including social insurance systems that can help farmers coping with drought episodes and building resilience to natural hazards.

Optimize virtual water flows

Whatever the means adopted to increase food production, water balance of Tunisia will remain in deficit and net Virtual Water (VW) flows will participate in a structural manner to national food security. However, a number of thoughtfully designed and implemented strategies would likely reduce the national water dependence and strengthen food security.

In general, Tunisia has interest to import the maximum of high water consuming agricultural products to compensate its internal deficits (in food and water) while exporting agricultural products with low water consumption and high added value. This orientation principle makes it possible to reduce the level of exploitation and degradation of internal resources while meeting the water and trade deficit. In this context, the only constraint is the excessive import of strategic commodities, those that constitute the basis of Tunisian food, particularly grains and edible oils. It is essential that a country can guarantee some independence in the production of staple foods to ensure at all times a minimum food security. But it is also important in this context to define the quantities of strategic food it is essential to locally produce at affordable prices. As for food grains, a comprehensive study should be undertaken to determine the "strategic" levels respectively for domestic production and import and their impact at various levels, including in particular those related to water resources. Indeed, all quantities of cereals produced or imported are not intended actually to basic supply and waste occurs at various levels of the value chain. Furthermore, and as for the management of water demand, it is essential to design and implement a policy of "food demand management", which aims to rationalize demand foodstuffs whose mass production is still closely linked the issue of water. The "regulation by the downstream" finds here all its interest.

9.3 Towards Green and Virtual Water Management

The challenge of water security involves many issues but number of water management problems are local in nature and are beyond the scope of uniform policies. Nonetheless, a central idea, essential in addressing National Water Security, is that a nation should cope with all its water resources within an integrate vision of all national development strategies. This holistic vision of the water resource allows for valorizing and optimizing all kind of water resources: improvement of the water use efficiency, valorization of rainfed agriculture, development of alternative water resources, optimization of the VW fluxes, etc.

The concept of WF developed over the last two decades allows for an analysis of international flows of VW. Because of its environmental significance, the assessment of the WF has become such an important issue that some countries like Spain have incorporated it into their legislation. From a quantitative standpoint, WF analysis demonstrates that the major share of water resources goes to the food production. This highlights the structural relationship between water, agriculture and food and leads to a fundamental rethinking and reinterpretation of the traditional water resource concept. For instance, the water embedded in the rainfed agriculture production and in the pastures takes part, as well as the water directly used for irrigation, in the food production.

The issue is not only how to develop all the factors and methods allowing access to a methodical knowledge of the GW and VW and their fluxes. It is also a matter of how to integrate their management by considering them as a resource on its own and thereby providing it with a legal basis and an institutional status. In that way, it will become possible to start talking about BW, GW and VW management (BGVWM) as the logical extension of the well established concept of integrated blue water resources management (IWRM). This will require a major change in the way the water issue is approached and addressed. Within this inclusive and comprehensive perspective, hydrologists, hydro-geologists, soil scientists and agronomists will need to broaden their focus beyond the classical approach of water and soil resources.

To give meaning to this new direction, the new Tunisian draft Water Code specifies four water resources categories: (i) surface water, (ii) groundwater, (iii) unconventional waters, (iv) agricultural soil water, stating that "agricultural soil water represents the infiltrated part of precipitations, which does not reach water table, is temporarily stored in agricultural soil and available either for direct evaporation in the case of a bare soil, or transpiration and plants consumption for a covered soil". The Code also stipulates that "the agricultural soil water remains subject to the legislation in force concerning soil and water conservation", and that "plans for water resources shall contain measures to encourage conservation and upgrading of soil water".

The "Water Dependency Index" (WDI) represents the ratio of net VW imports to the total water requirements for the food demand. The net VW imports are calculated as the difference between the total water requirements for food demand and the total WF of the food production. As the major part of water resources is directly or indirectly used in food production, the "Water Dependency Index" (WDI) measures the level to which a nation relies on virtual foreign water to ensure its food demand. Based on a comprehensive food-demand water balance model, the WDI formulation attempts to go beyond the appraisal of the water dependency level of nations to specify the balance sheet items related to the national food demand. The WDI, once implemented and correctly adjusted within a region or a country for which sufficient data are available, becomes a powerful instrument that makes it possible to prospect the future of the holistic water balance and to identify the impacts of the different water development programs on the water security.

In this regard, the prospective study for Tunisia at the horizon 2050 indicates that the improvement of "Water Dependency Index" (WDI) will depend on the capacity of the country to cope with all water resources and to improve food productivity either in the irrigated sector (BW) or in the rainfed farming one (GW). In particular the prospective study clearly shows that GW development can have significant positive effects on the food security. The relative increase of the GW is able to compensate the relative increase of the water requirements for food demand. Moreover, this index can be lowered if, in term of WF, the relative increase of the local agriculture production exceeds the relative increase of food demand. Better yet, the budget of food trade balance may positively improved by benefiting of comparative advantage associated with the exchange of VW. It appears thus that the developing rainfed agriculture is possible and its impact on water security is fundamental. This requires that we have to give to the rainfed agriculture the same level of interest that it has been given to the development of the irrigation, over the past four decades.

All these aspects will have far-reaching implications on the VW flows associated with international trade in food products, especially since a significant proportion of foodstuffs exchanges comes from rainfed agriculture. On the other hand, the long term National Water Security measured by the Water Dependency Index evolution, turns out to be extremely sensitive to the GW resources exploitation. Rainfed agriculture, without consequences on water resources, provides the largest part of foodstuffs production and plays an important role in the equilibrium of the food-stuffs trade balance.

These findings clearly indicate that the effective water management, in Tunisia and generally speaking in all water scarce countries, requires that all components of water availability and water demand are identified and accurately estimated considering climate change impacts. The improvement of the food security of a country expressed in terms of WDI will depend on the capacity of the country to improve food productivity either in the irrigated sector (BW, including non-conventional water resource) then in the rainfed crop production (GW, including pasturage). The methodology and analysis tools proposed here and applied to the water situation in Tunisia have proven to be a useful way to highlight some aspects of water issues and to address problems identifying possible solutions for future water management.

From this point of view, the WDI appears as major decision-making tool for sustainable water resources management, in particular in water scarce regions or countries. It is also a learning tool as well as a 'discussion-support' tool that provides a common platform for coherence of the activities of different actors and stakeholders. Furthermore, as it is based on a comprehensive food demand model, the WDI could be more explicitly detailed in order to bring out the different contributions to food production: "BW" referring to the use of ground and surface water as well as non-conventional water resource, "GW" referring to the water reserves of the soil and "VW" referring the flux of the "net VW import". The objective is to consider the extent to which greater value for all water resources could be achieved. To this regard, the WDI could be consolidated by financial indicators, for instance the coverage rate of the agrifood trade balance.

9.4 Tunisia: A Pragmatic Way for Addressing a Conceptual Model of the National Water Security

Tunisia water project has transformed the physical and social landscape of the country and sealed new inter-regional solidarities. But the almost total mobilization of water resources, the blue water, marks the end of an era: it shows how changing the water paradigm has become an urgent need and provides evidence for new policies adapted to the actual water opportunities and the water demands evolution. These policies that expand the water balance to all forms of resources, including water mobilized by rainfed crops, the green water, and the water equivalent of agrifood trade—the virtual water—are relevant when they are applied to water scarce countries. The Tunisian case study aims to achieve sustainable water security objectives at national level, and to bring about changes in the behavior of all actors with regard to water. This example provides a lot of information and experiences, which throw new light on the National Water Security issue. The case of Tunisia is an exemplary support to give information and instruments of general character, especially for all arid countries environments.

The Tunisian case provides a complete but still unfinished and imperfect conceptual model of the National Water Security paradigm with its multiple dimensions.

The first dimension is the historical heritage, and progress, embodied by a century of experiences, lessons, successes and also weaknesses and failures. A century, in which we saw the development of a modern era of hydraulics in the World and a social, demographic, economic, hydraulic and cognitive unprecedented growth in Tunisia. The National Water Security results from the heritage of this century marked by a permanent inventory of water resources and water knowledge acquisition, where Tunisia was a pioneer, by developing major water use regional master plans and many economic and social development plans, where hydraulics occupied a privileged place, with the realization of the essential facilities, mutations and reforms implemented to modernize the regulatory framework, protect water resources, control non-conventional sources, manage risks and anticipate conflicts of uses.

The second dimension is the best **knowledge of the national water balance**, which forms the basis of the Blue Water security dimension: implementation of the resources observation networks, knowledge of the natural rainfall resource, accurate assessment of surface runoff and groundwater flows, and their variability, knowledge of demand integrating the needs of society, agriculture and the environment. Knowledge of the hydrological context is essential: withdrawals; uses; regional imbalances and transfers; groundwater overexploitation; knowledge of anthropogenic water cycle; total water balance of Tunisia and the becoming of rainwater resource in all of its forms.

The third dimension: the new **paradigm of the National Water Security** is based on the extension of the national water balance assessment to two other fundamental components: (i) the green water linked to the rainfed agriculture and the extensive pasture localized in steppe regions, **the green water security dimension**, and (ii) the virtual water contained in the multilateral exchanges of agrifood products, **the virtual water security dimension**. The introduction of virtual water aims to optimize the holistic water balance of Tunisia and make sustainable, in the long term, the exploitation of all available water resources. The green water resources are not well known. Systematic green water development requires, the same way as conventional water resources, a careful assessment of its potential, a description of its spatial distribution and a characterisation of its temporal variability. The objective is to make this water potential more visible and more concrete in order to integrate it systematically into the prospects for development and exploitation of water and soil resources.

The fourth dimension represents **water demand management** associated with unconventional resources development, or the whole blue water efficiency dimension. This first considers aspects relating to conditions of water supply and management, as well as major challenges in key areas of water use in vital or priority sectors (drinking water needs), and the economic sector that consumes the most (agriculture). Then, possibilities and conditions for **use of non-conventional water**, like treated wastewater reuse, are considered as an extension to demand management policies.

The still debated issues—the not yet correctly resolved water problems, questions or concepts—**constitute the fifth dimension** of the National water security conceptual model. In spite of conceptual change in knowledge and understanding of the water security issue, very numerous water experts in arid countries continue to use, as water stress indicators, the Falkenmark thresholds, though based on the food self-sufficiency hypothesis. This translates well the necessity of building, and referring to, a complete conceptual model of the National Water Security. In the specific case of Tunisia, the water issues which are still being questioned lead to organize the debate around three major axes:

(a) **How to secure the blue water supply?** Better securing blue water supply to meet basic needs is relied on the next multiple aspects: (i) Looking for new ways to improve the national blue water budget. This implies the best knowledge of water resources systems, complete control of water use and groundwater overexploitation, update and adaptation of water institutions, rehabilitation/modernization of aging hydraulic infrastructures, finally considering all available opportunities to improve the national water budget; (ii) Reducing uncertainty margins on the water resources estimates, especially on: poorly concentrated surface water, transboundary flows, groundwater recharge, weakly renewable groundwater, recycled wadi floods, surface base flow subject to double counting, brackish groundwater; (iii) Transferring and underground storage of water surpluses from the wet regions in the North; (iv) Securing the drinking-water supply system; issues relating to water safety are: – lack of resources in the event of drought, circumstantial malfunctioning or floods causing supply disruptions, – risks of accidental pollution or poorly controlled contamination, – malicious acts, of criminal or terrorist nature, and situations of extreme social unrest or major conflict; (v) Safeguarding strategic reserves to secure water supply to metropolises in case of an ultimate crisis; (vi) Coping with groundwater overexploitation and the need for a new groundwater management mode involving all users; all aquifers should be regarded as strategic water reserves, with all that means, especially protection for sustainable use; (vii) Securing the water quality, which necessitates: – participation to the control of drinking-water quality, particularly by consumers, – systematic monitoring of water resources quality and pollution, – guarantee everywhere the good status of water; (viii) Other issues of political order which remain outstanding in the debate to combat poverty and social exclusion, like: – the almost systematic water transfer of poor inland areas to rich coastal regions; – improvement of drinking water quality service and sanitation in rural areas.

(b) **How does the water future appear?** The long-term water security covers the security of the water supply, but also food security since agriculture consumption accounts for a large share of water resources. How to meet the growing needs of the different water use sectors of the economy without impoverishing the irrigation sector? Without providing definitive answers to this question which has largely determined Tunisia's water policy, an outlook of water-balance evolution is elaborated in an overview that includes the balance sheets of the food balance. This way to deal with the long term National Water Security, which grants a prominent place to the water-food Nexus measured by the Water Dependency Index evolution, turns out to be extremely sensitive to the green water resources exploitation, and consequently to the climate change impact. In the term of this prospective exercise, the heavy trends, evolutions, and the major necessary reforms, in order to increase and reinforce the National Water Security, appear as follows: (i) Increase water efficiency and economic valorisation to ensure irrigation sustainability, (ii) Promote development and use of alternative water resources wherever possible, (iii) Implement a comprehensive strategy for the development of green water, (iv) Optimize virtual water flows; with an excessive import of strategic commodities, it is essential that the nation guarantee a minimum food security at all times; as in the case of water demand management, it is essential to design and implement a policy of "food demand management".

(c) **How to achieve good water governance?** This theme covers in a non exhaustive way legislative, institutional, regulatory and cognitive issues, most of which are subject to wide-ranging debates initiated some thirty years ago and formalized by the water community. Most of these issues are still open to debate. It is important for Tunisia to introduce management strategies that promote the environmental and heritage value of the resource by anticipating and controlling all forms of risks, to ensure the best possible use of the resource by preserving it and develop practical solutions for the promotion of all available water resources. All these objectives require in-depth reforms in the modes of governance and resource management: (i) institutional reforms: inter-sectoral coordination and improvements of decision-making mechanisms, local water management procedures, defining territorial unity of water management, strengthening and modernizing major public institutions to support water policies; (ii) legislative reform with the new Water Code; (iii) search for a water use optimization by integrating financial, economic and normative tools; (iv) improve intellectual, cognitive and scientific investments in the water field, and better use of water knowledge; (v) improve technical, scientific and institutional cooperation with neighboring countries for an effective and beneficial management of transboundary basins: support and strengthen success stories like the NWSAS Consultation Mechanism; use the lessons learned in the NWSAS to upgrade the cooperation level for the other transboundary basins.

9.5 As a Conclusion

The water vision which has so far formed the basis for Tunisian water policy appears to be reaching its limits of physical and economic effectiveness. The prospect of reallocating water available in agriculture sector places the water issue in the context of relationships with agricultural production, rural development and food security objectives. In the future, Tunisia needs to deal with all its water resources in order to ensure its water security and improve food security. Whether from surface or groundwater flow, stored in the soil or embedded in imported products, whether exploited by dams or water wells, captured by plants or produced by unconventional methods, any form of fresh water can help to increase resources and strengthen the potential of meeting basic needs. Blue water, green water, virtual water, all the resources are involved, in different ways, with National Water Security; in all cases the aim is to seek more and better exploitation procedures, to preserve and optimize their use.

The level of blue water exploitation is already very high and the current programs seem to further develop the hydraulic devices to retrieve surface water that still escapes mobilization. Beyond the debate over whether some of these programs have a regard for their financial implications and their environmental consequences, the national effort should first be devoted to maintaining and strengthening the potential of the existing infrastructure by combating all quantitative and qualitative degradation. Well preserved, water resources and soils must be exploited rationally to improve the efficiency of their use by strengthening demand management mechanisms: water saving, fight against waste, improvement of allocation and sharing procedures.

This is to place the objectives of resource protection, improving the quality of natural water and water-use rationalization at the heart of water-resource development and management structures. It will be necessary for this purpose to develop legislative tools and institutional arrangements needed to achieve these objectives. In particular, measures are required to provide a new impetus for water policy: recognition of the heritage value of water, organization of an institutional framework at geographic scale, compatible with the resource unity, introduction of the concept of good status of surface water and groundwater, enlargement of the maximum valuation principle to all kinds of water resources, strengthening the legislative framework for water planning, development of consultation mechanisms, establishment of arbitration and dispute-resolution processes. Adaptation of the legislative and regulatory framework is also needed in order to prepare human resources to support expected developments of larger-scale unconventional resources (desalination, treated wastewater reuse) that promise to be prominent in the medium term.

In Tunisia, a Mediterranean country, green water resources represent a considerable potential and rainfed agriculture holds a strong position in the food balance. Consolidation of food security cannot be conceived without decisive progress in valorization of green-water resources. This involves efforts at different levels in order to strengthen rainfed agriculture and improve yields of pasturelands: (i) At the scientific level, it is necessary to develop knowledge on the physical, physiological and genetic mechanisms of resistance of crops to drought, to increase storage capacity and water use; (ii) from the organizational and management standpoint, appropriate accompanying measures are required to better manage green water production variability: structural measures (construction of accompanying infrastructure: roads, grain silos, food reserves) and nonstructural (aid mechanism and support to farmers in dry years, insurance systems); (iii) from the legal point of view, the status of green water needs to be clearly defined: the principle of maximum enhancement of cubic meter of water, which is one of the main foundations of the Tunisian Water Code, should be extended to soil water resources. The ongoing review of the legal framework for water resources appears to be in line with this perspective.

The efforts aiming at increasing agricultural production are necessary but may not best serve the purpose of meeting the growing demand for food in quantity and quality. To offset the water deficit in terms of national agricultural production, notably during droughts, there will always be a need to use agrifood trade to balance supply and demand. The flow of virtual water, which appears on first analysis as an adjustment variable, must be optimized taking into account the limitations of internal water resources and the benefits of the comparative advantages that domestic agriculture can develop.

This holistic water vision leads to consider integrative Blue, Green and Virtual Water Management (BGVWM). To provide the basis for further progress in this direction and to sustain National Water Security, hydrologists, hydrogeologists, soil scientists and agronomists will need to broaden their focus beyond the classical approach of water and soil resources. It is also necessary to integrate water issues into all sectoral economic policies to promote all forms of available water resources and avoid unjustified exploitation or transfers and to define the objectives, orientations, rules and instruments of water policy on a regional scale and across the country. Improved yields of irrigation-water and drinking-water networks, the use of water-saving technologies, especially in irrigation, the adoption of appropriate systems of pricing and cost recovery, advocating water-saving user-awareness and sensitizing the general public to the preservation and appreciation of water, the full implementation of water-uses regulations, and control of hydrological risks, are the tools of a modern water policy.

Carrier of long-term strategies, water policy needs to be assimilated by all water stakeholders who adopt the objectives; it should also be known and understood by the public. The complexity of the actor system and the multitude of constraints reveal the need to establish effective coordination, which should bring the various stakeholders at national, regional and local levels to act together and harmonize their activities. Certainly, the integrated water-resource management cannot be conceived without the participation of water users; it is an accepted principle, an essential design, especially when dealing with limited resources subject to unsustainable exploitation. Popularization, information, communication, school education, then become essential components of management strategies. Scientific research, higher education, expertise and technology are closely related: theses strategies have to be accompanied by capacity-building programs, training and scientific education to meet the challenges facing the future of water.

Definitions and Terminology: Glossary of Key Terms

Aquifer

An aquifer is a geological formation where water is generally abundant and easy to extract

Shallow aquifer: aquifer closest to the ground also called **phreatic**, or **water table**
Deep aquifer: confined at its base and top by low permeability layers.
Renewable aquifer: aquifer regularly recharged by precipitation or runoff.
Weakly renewable aquifer: aquifer with large storage capacity and weak recharge rate.

Blue water

The share of precipitations that form fresh surface runoff and groundwater flow, whose sum, including exploitable weakly renewable water resource, is traditionally termed natural water resource.

Good status of water

Refers to the ecological status and bio-chemical status of surface water bodies and the quantitative status and chemical status of groundwater.

The **good ecological status** refers to a state in which all components of the ecosystem are active and preserved. As to the **good bio-chemical status** of surface water and groundwater, its definition is based on the identification of the various pollutants and the existence of quality standards.

The **good quantitative status** relates to groundwater. It is observed when water withdrawals remain below groundwater recharge capacities. Because abstractions and recharge are subject to considerable uncertainty, the quantitative status is generally assessed by analyzing multiannual water levels series.

M. Besbes et al., *National Water Security*,
https://doi.org/10.1007/978-3-319-75499-4

Green water

The share of average rainfall that does not runoff nor reach aquifers. It is temporarily stored in the soil and feeds "evapotranspiration", either for transpiration and plant consumption, which is the productive part of green water, or for direct evaporation from the soil for its non-productive part.

Hydraulic

Hydraulic structures: Water facilities built for water resources utilization or as protective structures against floods. Syn: **Hydraulic facilities, hydraulic works, hydraulic system**.
Hydraulic region: a hydrological basin including number of hydraulic systems and facilities.

Hydrological

Hydrological cycle, **hydrologic cycle**: Terrestrial water circulatory system showing the existing flows between major water reservoirs of the planet. Human activities impact the hydrological cycle, by their abstractions and discharges.
Hydrological balance: Evaluation of inputs, outputs and storage variations of a natural hydrological system over a period of time.

Mobilization

Action to collect a water resource, and make it available, generally by means of a hydraulic facility.

Mobilization rate: The share of water resources that is **mobilized**.

Overdraft

Aquifer overdraft is the amount of groundwater withdrawn from aquifer reserves. This is also termed aquifer depletion.

Overexploitation

Groundwater overexploitation refers to a state where, over a multi-year period, the abstraction from a given aquifer exceeds the recharge flux. When this imbalance persists for several decades, the exploitation regime is unsustainable, resulting in depletion of reserves and exhaustion of the aquifer.

Rainfall resource

Average total annual precipitation volume over a given region, also called rain water resource.

Rainfed water resource

Or **Rainfed crops rainfall resource**: means the rainfall supply of a given cultivated area.

Virtual water

Allan (1997) defined Virtual water as the water embedded in water-intensive commodities, circulating in the global trading system. This is termed also as the **virtual water flux**. The **Virtual water content** is the quantity of water effectively used in the production of a good or service, measured at the production site.

Water balance

Evaluation of water inputs, outputs and storage variations of a given system, whether natural or artificial, over a period of time. By default and without further specification, means the blue water balance.

Global water balance: water balance of the Earth. Syn. **World water balance**.
Water budget: same definition as water balance.
Holistic water balance: water balance relating to the complete water resources system (blue water, green water and virtual water) rather than to individual parts; also referred to as **comprehensive, complete**, **total**, **overal**, **whole**.

Water cycle

Same definition as the hydrological cycle, but can be applied to any system, whether natural or artificial. The circulatory system of water withdrawals, uses, consumptions, recycling and effluents in a given territory is referred to as the "**anthropogenic water cycle**".

Water dependency

For a given territory, represents the ratio of net virtual water volume (Imports - Exports) to the water equivalent of the total food demand.

Water equivalent

The same definition as for virtual water content and for water footprint of a product.

Water footprint

The water footprint (WF) of a product, measured in m^3/ton, is the volume of freshwater, blue and green, effectively used to produce the product. the WF or virtual water content of a product is measured at the place where it is produced.The water footprint of a given entity, individual, community, country, is defined as the total volume of freshwater used to produce the goods and services consumed by the entity (Hoekstra and Chapagain 2008).

Water resources

Natural water resources of a country: comprises the whole amount of freshwater (the blue water) effectively available in the country: (i) **internal renewable** water resources which include total average surface runoff and groundwater recharge produced by internal precipitation; (ii) total **external** water resources which covers transboundary net inflows from neighbouring countries; (iii) exploitable **weakly renewable** water resources of the country.

Exploitable water resources: The whole natural water resources of a country are not always exploitable. Exploitable water resources are water resources available for development, under given physical and chemical limits, economic conditions and predefined admissible environmental impacts.

Water resources exploitability: ability of natural water resources to be exploitable, within given limits.

Non conventional water resources: complementary quantities of water generated by desalination or wastewater recycling.

Renewable water resource: see renewable aquifer

Soil water resource: infiltrated part of rainfall temporarily stored in the ground, available either for direct evaporation from the soil (non-productive green water resource) or/and for plant evapotranspiration (productive green water resource).

Holistic water resources: Include (i) available water resources mobilized from surface water and groundwater, (ii) soil water resources, (iii) non conventional water resources, (iv) virtual water flows by trans-boundary trade.

Global water resources: fresh water resources of the Earth. Syn.: **World water resources**.

The manufacturer's authorised representative in the EU is Springer Nature Customer Service Centre GmbH, Europaplatz 3, 69115 Heidelberg, Germany. If you have any concerns regarding our products, please contact ProductSafety@springernature.com

Printed and bound by CPI Group (UK) Ltd, Croydon, CR0 4YY
05/10/2026
02228139-0011